Synthesis and Processing of Nanostructured Materials

Synthesis and Processing of Nanostructured Materials

A Collection of Papers Presented at the
29th and 30th International Conference on
Advanced Ceramics and Composites
January 2005 and 2006,
Cocoa Beach, Florida

Editor
William M. Mullins

General Editors
Andrew Wereszczak
Edgar Lara-Curzio

The
American
Ceramic
Society

BICENTENNIAL
1807
WILEY
2007
BICENTENNIAL

A JOHN WILEY & SONS, INC., PUBLICATION

Published by John Wiley & Sons, Inc., Hoboken, New Jersey
Published simultaneously in Canada.

For general information on our other products and services please contact our Customer Care Department within the U.S. at 877-762-2974, outside the U.S. at 317-572-3993 or fax 317-572-4002.

Wiley also publishes its books in a variety of electronic formats. Some content that appears in print, however, may not be available in electronic format.

Library of Congress Cataloging-in-Publication Data is available.

ISBN-13 978-0-470-08051-1
ISBN-10 0-470-08051-5

10 9 8 7 6 5 4 3 2 1

Contents

Preface

This proceedings contains a collection of papers submitted from the Functional Nanomaterial Systems Based on Ceramics focused session, held during the 29th International Conference and Exposition on Advanced Ceramics and Composites, January 23–28, 2005 and the Synthesis and Processing of Nanostructured Materials symposium, held during the 30th International Conference and Exposition on Advanced Ceramics and Composites, January 22–27, 2006. in Cocoa Beach, Florida.

Thanks and appreciation goes to those who attended and participated in these nanomaterial sessions, to the authors who submitted a paper for this volume, and to those who helped in the review process.

WILLIAM M. MULLINS

Introduction

This book is one of seven issues that comprise Volume 27 of the Ceramic Engineering & Science Proceedings (CESP). This volume contains manuscripts that were presented at the 30th International Conference on Advanced Ceramic and Composites (ICACC) held in Cocoa Beach, Florida January 22–27, 2006. This meeting, which has become the premier international forum for the dissemination of information pertaining to the processing, properties and behavior of structural and multifunctional ceramics and composites, emerging ceramic technologies and applications of engineering ceramics, was organized by the Engineering Ceramics Division (ECD) of The American Ceramic Society (ACerS) in collaboration with ACerS Nuclear and Environmental Technology Division (NETD).

The 30th ICACC attracted more than 900 scientists and engineers from 27 countries and was organized into the following seven symposia:

- Mechanical Properties and Performance of Engineering Ceramics and Composites
- Advanced Ceramic Coatings for Structural, Environmental and Functional Applications
- 3rd International Symposium for Solid Oxide Fuel Cells
- Ceramics in Nuclear and Alternative Energy Applications
- Bioceramics and Biocomposites
- Topics in Ceramic Armor
- Synthesis and Processing of Nanostructured Materials

The organization of the Cocoa Beach meeting and the publication of these proceedings were possible thanks to the tireless dedication of many ECD and NETD volunteers and the professional staff of The American Ceramic Society.

ANDREW A. WERESZCZAK
EDGAR LARA-CURZIO
General Editors

Oak Ridge, TN (July 2006)

NANOPARTICLE COLLOIDAL SUSPENSION OPTIMIZATION AND FREEZE-CAST FORMING

Kathy Lu, Chris S. Kessler
Department of Materials Science and Engineering
Virginia Polytechnic Institute and State University
211B Holden Hall-M/C 0237
Blacksburg, VA 24061

ABSTRACT

Nanoparticle suspension and forming are important areas. In this paper, the stability and rheology of Al_2O_3 nanoparticle suspensions at different dispersant concentration, suspension pH, and solids loading were studied. The most desirable suspension conditions were 7.5-9.5 for pH and 2.00-2.25 wt% of Al_2O_3 for poly(acrylic acid) (PAA) dispersant. 45.0 vol% Al_2O_3 solids loading can be achieved while maintaining good suspension flow for freeze casting. The maximum solids loading of the Al_2O_3 nanoparticle suspension was predicted to be 50.7 vol%. The preliminary results of the freeze-cast sample showed that suspension pre-rest before freezing was critical for achieving defect free microstructures.

INTRODUCTION

To overcome agglomeration and low packing density issues, colloidal processing is the preferred approach for nanoparticles. The unique ionic properties of water allow addition of ions to overcome the problematic particle-particle attraction; this ion-related stabilizing mechanism is called electrostatic stabilization. Another approach is to add a polymer dispersant to the suspension; polymer chains adsorb onto the particle surfaces and extend into water, and physically repel one another; this method of stabilization is called steric stabilization.[1] More effectively, an ionic polymer dispersant and the ionic properties of water can be used simultaneously to obtain a well-dispersed suspension; this method is called electrosteric stabilization.[2] A delicate balance must be maintained in the pH and the dispersant concentration. Inappropriate pH will result in particle attraction, not repulsion.[3] The adsorbed polymer must be thick enough to prevent close particle contact and counteract van der Waals forces. Too little polymer will cause bridging flocculation; too much polymer will cause depletion flocculation.[4,5,6,7] Ideally, the adsorbed polymer layer should be just thick enough to prevent van der Waals bonding. Since nanoparticles have much larger specific surface area than micron size particles, more polymer molecules will be adsorbed to cover the nanoparticle surface of the same mass; this will correspondingly reduce the solids loading of the nanoparticles even with ideal polymer layer thickness.

Viscosity measures the ability of the solid particles to flow relative to one another. A decrease in the particle size leads to suspensions of high viscosity because the effective solids loading is increased in the nanoparticle suspensions. The effective solids loading increase occurs because the particle diameter decrease creates large electrical double layer or adsorbed polymer layer volume that accounts as part of the effective solids loading. For high solids loading, the polymer concentration must be controlled with a much higher precision for the nanoparticle suspension than for the conventional suspensions.[8] The suspension must flow well into the mold and fill in the complex details or cavities of the mold. Intuitively, the higher the solid loading, the more viscous and difficult the flow will be. However, high solids loading is necessary to get a fully dense component. A balance must be achieved to address these opposite demands.

Freeze casting is a process that pours the suspension into a nonporous mold, freezes the suspension, demolds the component, and then dries the component under vacuum. Very desirably, freeze casting avoids defect formation by eliminating capillary force during drying and saves tremendous effort in binder removal. The technique is ideally suited for complex shapes and costs very little on tooling, which is a huge saving compared to dry pressing, isostatic pressing, or shock wave compaction.[9,10,11] Flexibility in forming complex shapes at minimal cost is also very suitable for rapid prototyping. Freeze casting also has the advantage of using less expensive and non-toxic dispersing medium, water.

The work reported here is focused on understanding PAA dispersant concentration, suspension pH, and Al_2O_3 solids loading effects on PAA adsorption onto Al_2O_3 nanoparticles and the stability and rheology of the nano-Al_2O_3 suspensions. By measuring the suspension rheology under different conditions, the maximum solids loading for the Al_2O_3 nanoparticle suspension is predicted. Based on the optimized dispersion conditions, the 40 vol% solids loading suspension has been freeze-cast. The preliminary characteristics of the freeze-cast components are reported.

EXPERIMENTAL PROCEDURE

Al_2O_3 nanoparticles with average particle size of 38 nm and specific surface area of 45 m^2/g were used in this study (Nanophase Technologies, Romeoville, IL). The Al_2O_3 was reported from the vendor to have 70:30 of δ: γ phases. Even though the average particle size was much less than 100 nm, the particle size distribution was wide and there was a small percent of large particles close to 100 nm. PAA (M_W 1,800, Aldrich, St Louis, MO) was used as a polymer dispersant with the polymer segment as [-$CH_2CH(CO_2H)$-].

For the preparation of the Al_2O_3 suspensions, 10 wt% glycerol ($C_3H_8O_3$, water basis, Fisher Chemicals, Fairlawn, NJ) was mixed with water and the mixture was homogenized for 5 min using a ball mill. Glycerol was used to lower the freezing temperature of the suspension and refine the ice microstructure for homogeneous component formation during freeze casting.[12] Al_2O_3 powder was added for a specific solids loading in 10 g increments along with an appropriate amount of PAA dispersant. Since low pH promotes PAA dispersant adsorption onto the nano-Al_2O_3 particles, HCl solution was added to lower the pH to 1.5.[13] The suspension was ball milled for 12 hrs with periodic adjustment of pH to 1.5. This procedure was used to make suspension of approximately 20 vol% Al_2O_3. NH_4OH was then used to adjust the suspension to the desired pH level. Depending on the final solids loading desired, nano-Al_2O_3 was added again in 10 g increments, along with the appropriate amount of PAA dispersant and the adjustment of the suspension pH. The suspension was then mixed for 24 hrs for complete homogenization. NH_4OH was again used to adjust the suspension to the desired pH level.

For suspension characterization, the pH of the suspensions was measured by a pH meter (Denver Instrument, Arvada, CO). Potentiometric titration was used to determine the amount of PAA dispersant adsorbed onto the Al_2O_3 particles in a suspension.[14,15] pH was adjusted to 9.5±0.05 before titrant HCl solution was added in order to promote the dissociation of PAA dispersant in water. Adsorption curve was developed for known PAA concentration blank solutions first. To measure the adsorption of PAA dispersant in an actual suspension, the suspensions with PAA dispersant and different solids loading were centrifuged at 2,500 rpm for 45 min before collecting the resulting supernatants. A known volume of the supernatant was titrated and the amount of un-adsorbed PAA dispersant was determined using the standard curve from the blank PAA solutions. The viscosities of the suspensions were measured by a rheometer with a cone-plate geometry (AR 2000, TA Instruments, New Castle, DE). All the suspensions were pre-sheared at 200 s^{-1} to impart similar shear history. The measurements were performed at equilibrium shear with controlled shear rates.

Slurries of 40 vol% solid loading were used for freeze casting. Freeze casting molds were developed using poly (dimethylsiloxane) epoxy (RTV 664, General Electric Company, Waterford, NY) and a fully dense flexural test specimen as a pattern. After curing, the flexural test specimen was removed, leaving a cavity in the epoxy. A transfer pipette was used to place the well dispersed Al_2O_3 suspension into the mold to reduce air bubbles and promote filling. The Al_2O_3 suspension filled epoxy mold was placed in a freeze dryer (Labconco Stoppering Tray Dryer, Labconco, Kansas City, MO) and cooled to -35° C for 2 hrs. The samples were demolded after freezing and then exposed to low pressure (<10 x 10^3 Pa) for 36 hrs to allow the ice to sublimate. The microstructures of the freeze-cast samples were examined by a LEO 1550 field emission scanning electron microscope (Carl Zeiss MicroImaging, Inc, Thornwood, NY).

RESULTS/DISCUSSION
1) PAA Adsorption and Suspension Rheology
    Hydrogen bonding is a ubiquitous mechanism for polymer adsorption onto hydrophilic surfaces. Mathur et al. studied the nature of oxide surface hydroxyls and indicated that isolated surface hydroxyls constitute the surface of Al_2O_3.[16] Based on this understanding, PAA adsorption onto Al_2O_3 is expected to be affected by the PAA concentration. However, how such adsorption mechanism plays a role in high solids loading suspensions has not been examined. To study the PAA concentration effect on its adsorption onto Al_2O_3 nanoparticles in a high solids loading suspension, the solids loading of the suspension was fixed at 30 vol% while the PAA concentration was varied between 1.00-2.50 wt% of Al_2O_3 as shown in Fig. 1. As the PAA concentration increases up to 2.00 wt% of Al_2O_3 PAA concentration, the adsorbed PAA amount onto Al_2O_3 particles also increases. As more PAA is added to the suspension, the adsorbed PAA amount reaches a plateau at 0.31 mg/m^2. The adsorption plateau represents a saturation adsorption level at which the Al_2O_3 particle surfaces are fully covered by PAA dispersant. For the studied system, the minimum PAA concentration needed is 2.00 wt% of Al_2O_3. On the other hand, the relative amount of PAA adsorbed onto the Al_2O_3 nanoparticles decreases monotonically as more PAA is added, due to the blocking of some active sites from the adsorbed PAA polymer onto the solid surface; this means free PAA concentration increases along with total PAA concentration increase. After the adsorbed PAA amount reaches the saturation plateau for monolayer coverage, additional PAA dispersant only exists as free polymer in the suspension, resulting in 'overdispersing'.

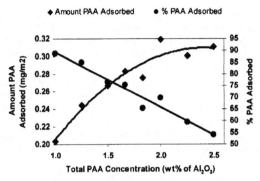

Fig. 1. PAA dispersant adsorbed amount and the adsorbed relative percent vs. the total PAA concentration.

The viscosity measurements at different shear rates for the suspensions of 30% solids loading and different PAA concentrations are shown in Fig. 2. The suspension has the lowest viscosity at 2.00-2.25 wt% of Al_2O_3 PAA concentration. As the PAA content increases or decreases, the viscosity increases. The viscosity increase at low PAA concentration can be explained by a shortage of PAA adsorption onto Al_2O_3 (incomplete PAA coverage of Al_2O_3) and the resultant bridging flocculation. The viscosity increase at high PAA concentration can be explained by the presence of free polymer in the suspension--depletion flocculation. A balance needs to be maintained in the PAA adsorbed onto the nano-Al_2O_3 particles and the free PAA dispersant in the suspension. The results in Fig. 2 serve as an important guide in optimizing the dispersion conditions. For the studied system, the optimal PAA content is 2.00-2.25 wt% of Al_2O_3.

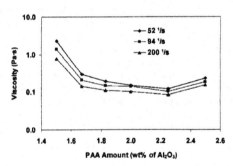

Fig. 2. PAA concentration effect on the suspension viscosity at 30 vol% Al_2O_3 solids loading.

Based on the PAA concentration effect on the suspension stability, PAA concentration of 2.00 wt% of Al_2O_3 was used for the pH effect study while the solids loading was kept at 30 vol%. pH was studied from 5.5 to 10.5. As shown in Fig. 3, PAA has the highest adsorption at pH 5.5. This is consistent with the theory that low pH promotes PAA adsorption onto Al_2O_3

nanoparticle surfaces. As pH increases, both PAA adsorption concentration and the relative amount of the adsorbed PAA decrease. From pH 5.5 to 10.5, the PAA adsorption concentration decreases from 0.44 mg/m^2 to 0.27 mg/m^2, a 38.6% decrease, while the adsorbed PAA percent decreases from 92.0% to 57.5%, a relative 37.5% decrease.

It should be noted that the PAA adsorption concentration is only one aspect of the polymer stabilization mechanism. PAA conformation on the Al_2O_3 nanoparticle surface also plays important roles. At low pHs, PAA tends to have more train configuration on the Al_2O_3 surfaces and hinders suspension stabilization. At high pHs, PAA has more loop and tail configurations and provides electrosteric stabilization. When the pH value is too high, PAA adsorption onto Al_2O_3 particles is too low, depletion flocculation results and viscosity increases. The viscosity measurement (Fig. 4) shows that at pH 5.5, the suspension is very viscous and there is substantial shear thinning. At pH 10.5, there is also noticeable shear thinning and high viscosity. At pH 7.5-9.5, the suspension has low viscosity and is the most desirable pH range for stabilizing the Al_2O_3 nanoparticles.

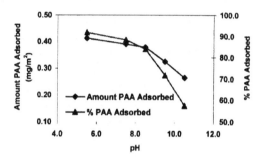

Fig. 3. Suspension pH effect on PAA adsorption at 30 vol% solids loading and 2.0 wt% of Al_2O_3 PAA concentration.

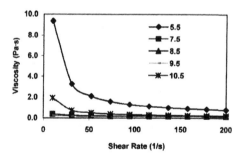

Fig. 4. Viscosity change vs. shear rate for the 30 vol% solids loading and 2.0 wt% of Al_2O_3 PAA concentration suspensions at different pH values.

For the solids loading effect study, all the suspensions were controlled at 2.00 wt% of Al_2O_3 for the PAA concentration and 9.5 for the final pH. Fig. 5 shows that as the Al_2O_3 solids loading increases from 20.0 to 45.0 vol%, the PAA adsorbed amount and the relative amount of PAA adsorbed onto Al_2O_3 nanoparticles increase. Since most of the prior research was focused on very dilute systems (< 20 vol%), this observation revealed trend that has not been detailed before. The reason can be traced back to the more favorable PAA adsorption onto the Al_2O_3 surfaces at lower pH. During the Al_2O_3 solids loading adjustment, Al_2O_3 and PAA additions shift the suspension pH to more acidic conditions from pH 9.5. This suspension adjustment process promotes the PAA adsorption onto Al_2O_3 nanoparticles as demonstrated in Fig. 3. After the PAA adsorption at the more acidic condition, the final adjustment of pH to 9.5 does not change the bonding between PAA and Al_2O_3 and the higher PAA adsorption. From 30.0 to 45.0 vol% Al_2O_3 solids loading, the PAA adsorption difference is similar to the adsorption difference for a pH decrease from 9.5 to 8.5 for the 30 vol% solids loading suspension. This solids loading and PAA adjustment effect can probably be minimized by controlling pH at 9.5 with very small increment of Al_2O_3 and PAA additions. However, this is not very practical since the suspension needs to be mixed thoroughly for homogenization at each small addition. Instead, Fig. 5 provides practical information for the optimal dispersion conditions at different Al_2O_3 solids loading levels.

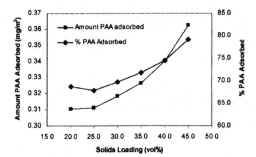

Fig. 5. Solids loading effect on adsorbed PAA amount and relative percent of PAA adsorbed onto Al_2O_3 nanoparticles.

Fig. 6 shows that the suspensions have reasonably low viscosity and are well dispersed from 20.0-45.0 vol% solids loading, consistent with the prior measurement that the suspensions have about -40.0 mV zeta-potential.[17] It can also be easily observed that the viscosity is a strong function of Al_2O_3 solids loading. With the solids loading increase, the viscosity increases. As the shear rate increases, the viscosity decreases monotonically. The viscosity becomes more sensitive to the Al_2O_3 solids loading increase at high solids loading levels.

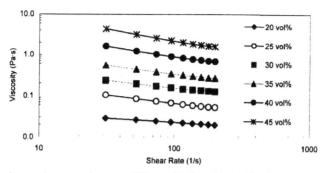

Fig. 6. Viscosity vs. shear rate change at different Al_2O_3 solids loading levels.

2) Maximum Solids Loading Prediction

Nano-Al_2O_3 has large specific surface area that adsorbs large amount of dispersant PAA, which can quickly reduce the particle packing efficiency. When excessive PAA dispersant is used, some PAA polymer chains stay freely in the suspension, further decreasing the achievable maximum solids loading. Mathematical models have been developed to relate the amount of the solids loading with the viscosity. Among these is the semi-empirical Krieger-Dougherty equation:

$$\eta_r = \frac{\eta}{\eta_0} = \left(1 - \frac{\phi_{eff}}{\phi_{max}}\right)^{-n\phi_{max}} \qquad (1)$$

where η_r is the relative viscosity, η is the suspension viscosity, η_0 represents the viscosity of the suspension medium (water), and $\dfrac{\phi_{eff}}{\phi_{max}}$ represents the solids loading over the maximum solids loading (~0.64). Here, n is a constant representing the intrinsic viscosity of the suspension.[18] Liu showed that most of the equations describing particle packing-viscosity behavior at high strain rate (~100 s^{-1}) follow the simple form:

$$1 - \eta_r^{-\frac{1}{n}} = a \cdot \phi + b \qquad (2)$$

where a and b are constants. Assuming that the Al_2O_3 nanoparticle suspensions follow the model, the viscosity of the suspensions under different solid loading levels can be used to extrapolate the theoretical maximum solids loading, at the point where viscosity is infinite.[19] In this study, the maximum solids loading at pH 9.5 and PAA concentration 2.00 wt% Al_2O_3 can be predicted based on the viscosity at lower solids loading levels, which is 50.7 vol% as shown in Fig. 7 (it is assumed n=2 here as in most models). In this study, up to 45.0 vol% Al_2O_3 solids loading has been achieved with good suspension flow, which is even higher than that of some dry compaction processes.

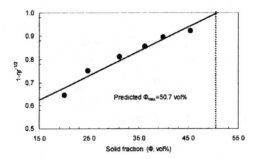

Fig. 7. Maximum solids loading prediction for the Al_2O_3 nanoparticle suspensions when PAA is at 2.00 wt% Al_2O_3, pH is at 9.5, and strain rate is at 94.0 s^{-1}.

3) Freeze Casting

In this study, the 40.0 vol% solids loading Al_2O_3 suspension was freeze cast. It was observed that the pre-rest of the filled suspension has significant effect on the freeze cast sample quality. After mold filling. the trapped air bubbles should be allowed to escape from the suspension under zero shear stress. This is necessary for the studied system due to the high solids loading. The fracture surfaces of the samples under different pre-freezing treatment are shown in Fig. 8 at different magnifications. All the samples were frozen under the same conditions as outlined in the experimental section. As Fig. 8 shows, there is no cracking for the sample with pre-rest (a) but extensive cracking for the sample without pre-rest (b) at low magnification. At high magnification, the sample with pre-rest shows more flat fracture surface and uniform structures. For the sample without pre-rest, very rough fracture surface and micro-voids can be easily seen. Further work is needed to optimize the freeze casting conditions, evaluate the macroscopic properties of the sample, and study the fundamental freeze-casting mechanisms.

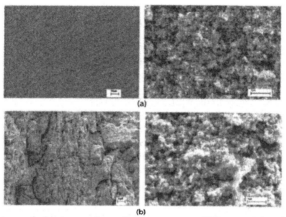

Fig. 8. Freeze cast sample fracture surface microstructures at different pre-freezing conditions: (a) with pre-rest, (b) without pre-rest.

CONCLUSIONS

PAA dispersant concentration, suspension pH, and Al_2O_3 solids loading effects on PAA adsorption onto Al_2O_3 nanoparticles were studied; the stability and rheology of the nano-Al_2O_3 suspensions under different levels of the above three parameters were examined. The most desirable suspension conditions were identified to be 7.5-9.5 pH and 2.00-2.25 wt% of Al_2O_3 PAA concentration; and the highest Al_2O_3 solids loading can be achieved at more than 45.0 vol%. The maximum solids loading of the nano-Al_2O_3 suspension was predicted to be 50.7 vol%. Pre-rest of the suspension was vital for achieving desirable solid samples.

ACKNOWLEDGMENTS
The authors are grateful for the partial funding support of Oak Ridge Associated Universities.

REFERENCES
[1]J. A. Lewis, "Colloidal Processing of Ceramics." *J. Am. Ceram. Soc.*, **83** [10], 2341-2359 (2000).
[2]J. Cesarano III, I. A. Aksay, and A. Bleier, "Stability of Aqueous Alpha-alumina Suspensions with Poly(Methacrylic Acid) Polyelectrolyte," *J. Am. Ceram. Soc.*, **71** [4], 250-255 (1988).
[3]J. M. Cho and F. Dogan, "Colloidal Processing of Lead Lanthanum Zirconate Titanate Ceramics," *J. Mater. Sci.*, **36** [10], 2397-2403 (2001).
[4]D. Napper, *Polymeric Stabilization of Colloidal Dispersions*; pp. 8-15, Academic Press, London, United Kingdom, 1983.
[5]W. M. Sigmund, N. S. Bell, and L. Bergstrom, "Novel Powder-Processing Methods for Advanced Ceramics," *J. Am. Ceram. Soc.*, **83** [7], 1557-1574 (2000).
[6]A. L. Ogden and J. A. Lewis, "Effect of Nonadsorbed Polymer on the Stability of Weakly Flocculated Suspensions," *Langmuir*, **12** [14], 3413-3424 (1996).

[7]Q. Li and J. A. Lewis, "Nanoparticle Inks for Directed Assembly of Three-Dimensional Periodic Structures," *Adv. Mater.*, **15** [19], 1639-1643 (2003).

[8]A. Dietrich and A. Neubrand, "Effects of Particle Size and Molecular Weight of Polyethylenimine on Properties of Nanoparticulate Silicon Dispersions," *J. Am. Ceram. Soc.*, **84** [4], 806-12 (2001).

[9]M. Torkar, V. Leskovsek, B. Sustarsic, and P. Panjan, "Failure of Tools for Metallic Powder Compaction," *Eng. Failure Analysis*, **9** [2], 213-219 (2002).

[10]H. Kamiya, K. Isomura, and G. Jimbo, "Powder Processing for the Fabrication of Si_3N_4 Ceramics, 1. Influence of Spray-Dried Granule Strength on Pore-Size Distribution in Green Compacts," *J. Am. Ceram. Soc.*, **78** [1], 49-57 (1995).

[11]B. Jodoin, "Effects of Shock Waves on Impact Velocity of Cold Spray Particles," pp. 399-407 in *Thermal Spray 2001: New Surfaces for a New Millennium*. Edited by C. C. Berndt, K. A. Khor, and E. F. Lugscheider. ASM International, Materials Park, Ohio, 2001.

[12]S. W. Sofie and F. Dogan, "Freeze Casting of Aqueous Alumina Slurries with Glycerol," *J. Am. Ceram. Soc.*, **84** [7], 1459-1464 (2001).

[13]J. Cesarano and I. A. Aksay, 'Processing of Highly Concentrated Aqueous Alpha-Alumina Suspensions Stabilized with Poly-Electrolytes,' *J. Am. Ceram. Soc.*, **71** [12], 1062-1067 (1988).

[14]R. Arnold and J. T. G. Overbeek, "The Dissociation and Specific Viscosity of Polymethacrylic Acid," *Recueil*, **69**, 192-206 (1950).

[15]Y. Q. Liu and L. A. Gao, "Dispersion of Aqueous Alumina Suspensions Using Copolymers with Synergistic Functional Groups," *Mater. Chem. Phys.*, **82** [2], 362-369 (2003).

[16]S. Mathur and B. M. Moudgil, 'Adsorption Mechanism(s) of Poly(Ethylene Oxide) on Oxide Surfaces,' *J. Colloid Interface Sci.*, **196** [1], 92-98 (1997).

[17] K. Lu, C. S. Kessler, "Colloidal Dispersion and Rheology Study of Nanoparticles," *J. Mater. Sci.*, in press.

[18] I. M. Krieger and M. Dougherty, "A Mechanism for Non-Newtonian Flow in Suspensions of Rigid Spheres," *Trans. Soc. Rheol.*, **3**, 137-152 (1959).

[19]D. Liu, "Particle Packing and Rheological Property of Highly-Concentrated Ceramic Suspensions: Φm Determination and Viscosity Prediction" *J. Mat. Sci.*, **35** [21], 5503-5507 (2000).

SYNTHESIS, CHARACTERIZATION AND MEASUREMENTS OF ELECTRICAL PROPERTIES OF ALUMINA-TITANIA NANO-COMPOSITES

Vikas Somani and Samar J. Kalita

Department of Mechanical, Materials and Aerospace Engineering

University of Central Florida

Orlando, FL 32816-2450

ABSTRACT

In this study, we have synthesized and characterized nanocrystalline alumina-titania (Al_2O_3-TiO_2) composites *via* a simple sol-gel process, using aluminum propoxide and titanium propoxide as precursor chemicals. Propanol and 2-methoxy ethanol were used as solvent and stabilizer, respectively. The as-formed gel was heat treated at 400°C, to obtain amorphous powder. The amorphous powder was subsequently calcined at 700°C and 900°C. Phase evolution, phase composition, crystal structure and crystallite-size of the synthesized powder were determined using X-ray diffraction (XRD) technique. Crystallite-size was further confirmed by high-resolution transmission electron microscopy (HR-TEM). HR-TEM results of powder calcined at 700°C showed agglomerates of powder particles, with particle-size in 15 - 20 nm range. The synthesized powder was uniaxially pressed using a steel mold and then sintered at elevated temperature (1000-1500°C) for densification study and electrical property measurements. XRD technique was used to study phase composition of the sintered pellets. Dielectric constant and dissipation factor of the sintered pellets were measured. The effects of sintering time, temperature and phases present on the electrical properties of the sintered pellets, were studied.

INTRODUCTION

Al_2O_3-TiO_2 systems are promising candidates for use in electronic devices as dielectrics. SiO_2, the most widely used dielectric till date, is reaching its physical limits due to miniaturization. It has been shown that SiO_2 films with thickness less than 7 nm lead to significant increase in leakage current. This drives the need to find new dielectrics. A number of reports in the last few years are based on potential dielectric materials like HfO_2, Al_2O_3, ZrO_2, Ta_2O_5, TiO_2, $BaTiO_3$ and SiN.[1] In particular, HfO_2 and Al_2O_3 with dielectric constants (κ) of 25 and 9, respectively, are believed to be materials of choice in near-future applications. Recently, there has also been an upsurge in research of compound dielectric materials, for example $ZrSiO_4$ and $HfSiO_4$.[2]

Alumina (Al_2O_3) has excellent insulating properties ($\kappa \sim 9$), a high band gap ~ 9.9 eV, and is resistant to most chemicals. While, titania (TiO_2) possesses a much higher $\kappa \sim 80\text{-}120$ depending on its crystal structure.[3] The disadvantage with Al_2O_3 is its relatively low dielectric constant, which limits its utility and life-span to only a few years. TiO_2, on the other hand, is unstable on silicon resulting in high leakage current. Aluminum titanate (Al_2TiO_5, AT), a compound of Al_2O_3 and TiO_2 is known to have excellent thermal shock resistance, low thermal expansion coefficient (0.2×10^{-6} to 1.0×10^{-6}) and high melting point.[4, 5] AT is usually synthesized by solid state reaction between equimolar mixture of Al_2O_3 and TiO_2 powders, at temperatures higher than 1400°C.[6, 7] The endothermic reaction can be written as:

$$Al_2O_3 + TiO_2 \rightarrow Al_2TiO_5$$

AT dissociates to corundum and rutile in the temperature range of 750°C-1280°C, and to overcome this demerit, stabilizers like Fe_2O_3, MgO or SiO_2 are added.

It has been established that nanocrystalline powder of various ceramic materials show enhanced sintering behavior, better mechanical and electrical properties.[8] Sol-gel processing technique has been successfully exploited to synthesize nanopowder and thin films. However, only a few studies have been reported on sol-gel synthesis of aluminum titanate, as well as alumina-titania nanocomposites. We hypothesized that nanocomposite systems of Al_2O_3 and TiO_2 would help us unite the merits of both Al_2O_3 and TiO_2 and possibly develop new material systems for future dielectrics. With this hypothesis, we synthesized of Al_2O_3-TiO_2 nanocomposites by sol-gel technique; and systematically studied their sintering and densification behavior along with their dielectric properties.

EXPERIMENTAL PROCEDURE

Powder Synthesis

Al_2O_3-TiO_2 nanocomposite powder was synthesized using aluminum isopropoxide (reagent grade, Fisher Scientific, USA) and titanium (IV) isopropoxide (purity 98+%, Fisher Scientific, USA) as precursor chemicals and propanol (1 N, Alfa Aesar, Ward Hill, MA) as the solvent. Hydrochloric acid (HCl) was used to control the pH of the solution. In order to prepare the sol 20 ml of propanol was heated in a beaker to 50°C. Subsequently, stoichiometric amount of aluminum isopropoxide was added into hot propanol and refluxed at 120°C, for 2 h. The refluxed solution was allowed to cool in air. In another beaker, stoichiometric amount of titanium (IV) isopropoxide was dissolved in 5 ml propanol. The as-formed aluminum isopropoxide solution was then added drop-wise into the solution of titanium (IV) isopropoxide in propanol using a burette. The resultant solution was continuously stirred, during the entire process. Hydrochloric acid was added during this process to maintain the pH of the solution between 3.5 and 7.0. As-produced Al-Ti sols at different pH were refluxed at 120°C, for 2 h. The resulting solutions were filtered and then dried in a muffle furnace to obtain amorphous powder. The as-formed amorphous powders were calcined at 700°C for 2 h. Based on XRD phase analysis of these powders, new sol was prepared with pH 5.5 only. Amorphous powder obtained from this sol was calcined separately in batches at different temperatures, viz., 400°C, 700°C and 900°C in a muffle furnace under atmospheric conditions, for crystallization. A flowchart of the powder synthesis process is shown in figure 1.

Powder Characterization

X-ray diffraction (XRD) technique was used to study the phase evolution and phase transformation during synthesis as a function of solution pH and calcination temperature. During initial phase of this work, agglomerated powder obtained from solutions having different pH (3.5, 5.5 and 7.0), calcined at 700°C for 2 h, were ground into a fine powder, by mortar and pestle and analyzed for phases. The results showed evolution of crystalline phases in powder synthesized using solution with pH of 5.5. Hence, we focused on studying the characteristics of powder synthesized, using solution with pH of 5.5 only. Powder agglomerates obtained after calcination at 400°C, 700°C and 900°C for 2 h, in separate batches, were ground into a fine powder using mortar and pestle, and analyzed for phase composition. A Rigaku diffractometer (Model D/MAX-B, Rigaku Co., Tokyo, Japan) equipped with Ni filtered CuKα radiation (λ =0.1542 nm) at 40 kV and 40 mA settings was employed for the purpose. Based on XRD analysis, powder calcined at 900°C for 2 h was examined for morphology, grain size and lattice fringes using a HR-TEM (Model Tecnai – Philips F30, FEI Co., Hillsboro, OR).

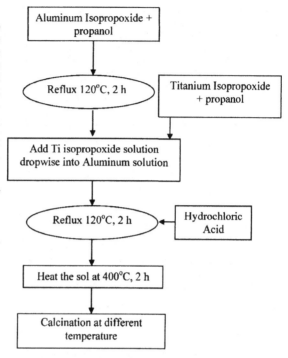

Figure 1. Flowchart showing steps involved in the synthesis of alumina-titania composite nano-powder.

Sintering and Densification Study

As-synthesized alumina-titania composite nanopowder calcined at 700°C was compacted in a steel mold, using a uniaxial single action manual hydraulic press (Carvar Press Inc, USA). These green structures were sintered at elevated temperatures (1000, 1100, 1200, 1300, 1400 and 1500°C), to investigate their sintering behavior, densification characteristics and phase transformation as a result of sintering. Some of the sintered structures were also characterized for electrical properties. During the preparation of the green specimens, a dry P.T.F.E film (made with Dupont Krytox) was sprayed on the inside surface of the steel mold and the punch, to lubricate and reduce friction. Uniaxially compacted green pellets with average dimensions of 12.5 mm in diameter and 3.0 mm in thickness, were sintered in a high temperature programmable muffle furnace in ambient atmosphere, and at different temperatures for 2 h. A heat treatment cycle was developed to allow better densification and to avoid cracking. These pellets were initially heated to 400°C at a heating rate of 8 °C/min and homogenized at this temperature, for 30 min. In the second step, temperature was increased to the desired final level; at a slow heating rate of 4°C/min. Pellets were subsequently cooled to room temperature, at 10°C/min. Three specimens were heat-treated at each sintering temperature and the values here reported, are the average of the three samples. Geometric bulk density (ρ_g) of each pellet was evaluated from the measurements of the mass of specimen and its volume (determined by dimensional measurements) using equation 1.

$$\text{Geometric bulk density } (\rho_g) = \text{Mass / Volume} \qquad (1)$$

Linear and volumetric shrinkages associated with sintering at different temperatures were also calculated from geometric measurements of green and sintered specimens.

Phase Analysis of the Sintered Specimens

XRD technique was used to study the effect of sintering temperature on phase purity and phase transformation of the alumina-titania nano-composites. Ceramic structures sintered at different temperatures were ground into a fine powder, using mortar and pestle. The fine powder thus obtained, was used for XRD analysis in a Rigaku diffractometer, as stated under *Powder Characterization*.

Electrical Characterization

Alumina-titania nano-composite structures sintered at 1000°C, 1200°C and 1400°C for 2 h, were used to measure electrical properties. Silver paste (electrode) was applied on both sides of these sintered pellets and the resulting structures were dried at 400°C for 30 min, in a muffle furnace, in air. The dielectric constant and dissipation factor of these sintered pellets were measured at 1 MHz, using a precision LCR meter from Agilent technology (Model 8284A) and 16451 B dielectric test fixtures. The effects of phase composition, density and porosity of the sintered structures on the electrical properties were investigated.

RESULTS AND DISCUSSIONS

Figure 1 is a flowchart, showing the steps in the synthesis of alumina-titania composite nano-powder. During the synthesis process, it was interesting to observe the changes of color in the amorphous and the calcined powders. Amorphous powder obtained from sols, maintained at all three pH *viz.*, 3.5, 5.5 and 7, were initially light brown in color. After calcination, the color of the powder from the sol of pH 5.5 changed into white, while no color change was observed in case of powders from the other two sols (pH 3.5 and 7.0).

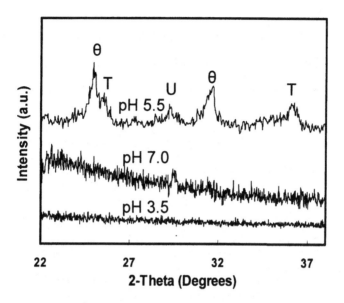

Figure 2. XRD patterns of powder synthesized from solution maintained at different pH. Calcination was done at 700°C for 2 h. The phases present were θ-Al₂O₃ and anatase TiO₂ with an unknown peak, which are marked as 1, 2 and U, respectively.

The powder synthesized from solution maintained at three different pH values, *viz.*, 3.5, 5.5 and 7.0 were analyzed, using XRD (figure 2). No distinct peaks were observed in the XRD pattern of powder synthesized from sol of pH 3.5. This XRD pattern confirms that the synthesized powder was amorphous in nature. XRD pattern of powder synthesized from sol of pH 7.0 shows a few low intensity broad peaks, confirming that it is not an optimum pH to obtain crystalline powder. On the other hand, XRD pattern of powder synthesized from sol of pH 5.5, showed clear broad peaks of θ-Al₂O₃ and anatase TiO₂. From this XRD pattern, it is clear that the powder is crystalline and was of a very fine size. It is well known that the reaction rate of molecular precursors used in the sol-gel process, is highly dependent on the processing variables like pH, concentration, solvent and temperature. In this process, we fixed the solvent and concentration but altered the pH and temperature, to find the optimum synthesis condition. The formation of α-Al₂O₃ from γ- Al₂O₃ is known to occur, by a series of polymorphic transformation on heating.[9]

$$\gamma\text{-}Al_2O_3 \xrightarrow{800} \delta\text{-}Al_2O_3 \xrightarrow{900} \theta\text{-}Al_2O_3 \xrightarrow{1100} \alpha\text{-}Al_2O_3$$

Here, it can be observed that θ-Al₂O₃ is stable at relatively low temperatures. This is due to the presence of TiO₂ and structural changes, forced as an effect of the nanosize of the particles. XRD results showed that variations in pH do play a role in the synthesis. These XRD patterns also revealed that sol having a pH of 5.5 was good for the synthesis of alumina-titania composite powder. Therefore, in the next stage of our research, we studied the phase evolution as a function of calcination temperature, in powders synthesized using sol with pH of 5.5. The powders were calcined at 400°C, 700°C and 900°C, for 2 h. XRD pattern of the powder calcined at 900°C is shown in Figure 3.

Figure 4a shows the TEM micrograph of the powder calcined at 700°C, for 2 h. Although, the calcined powder was ground manually using a mortar and pestle they are in the form of agglomerates. These agglomerates were found to be 0.3 to 0.6 μm in size. The agglomerates in turn, are formed of very fine particles, of about 15 - 20 nm in size. A HR-TEM micrograph of the particle is shown in figure 4b. The lattice fringes and ordered arrangement of atoms is apparent which, confirms the crystalline nature of the particles.

Uniaxially compacted alumina-titania nanocomposite pellets were sintered at 1000°C, 1200°C, 1300°C, 1400°C and 1500°C, for 2 h. Figure 5a is the plot of variation in sintered density, with increase in the sintering temperature. Volume shrinkage and linear shrinkage were calculated from the difference in the geometric measurements between green and sintered pellets and the calculated results were plotted as a function of sintering temperature, as shown in figures 5b and 5c, respectively. It is clear from these plots that geometric sintered density increased sharply from 1.39 g/cm³ at 1000°C to 2.15 g/cm³ at 1200°C. The corresponding volume shrinkage in this range almost doubled from 24% to 46%. The average sintered densities of the pellets, sintered at 1200°C

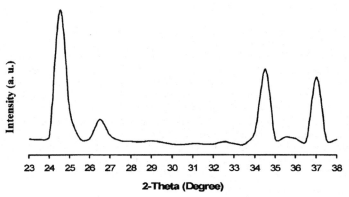

Figure 3. XRD pattern of powder calcined at 900°C.

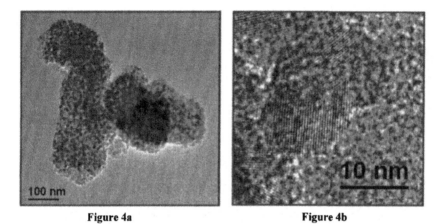

Figure 4a **Figure 4b**

Figure 4. (a) TEM micrograph of Al_2O_3-TiO_2 composite nanopowder agglomerate calcined at 700°C for 2 h. (b) HR-TEM micrograph of the nanopowder showing the lattice fringes.

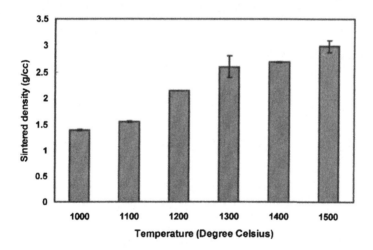

Figure 5a. Plot showing variation in sintered density of Al_2O_3 - TiO_2 composite pellets sintered at 1000°C, 1200°C, 1300°C, 1400°C and 1500°C in furnace atmosphere.

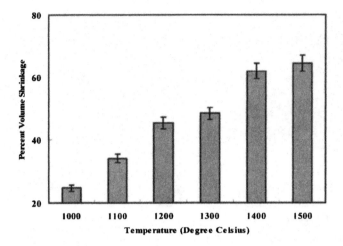

Figure 5b. Plot showing variation volume shrinkage of Al_2O_3 - TiO_2 composite pellets sintered at various temperatures in furnace atmosphere.

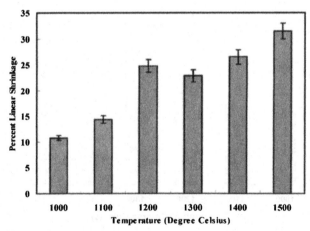

Figure 5c. Plot showing variation in linear shrinkage of Al_2O_3- TiO_2 composite pellets sintered at various temperatures in furnace atmosphere.

and 1300°C were found to be almost constant; though, the volume shrinkage increased from 45% to 56%. Thereafter, the sintered density increased to 2.6 g/cm³ in pellets sintered at 1400°C and the corresponding volume shrinkage at this temperature was found to be 62%. The highest sintered density of 2.99 g/cm³ is achieved at 1500°C, corresponding to volume shrinkage of almost 65%. The highest sintered density achieved is 80.7% of theoretical density of Al_2TiO_5. As expected, an increase in sintering temperature led to an increase in sintered density. The linear shrinkage of the pellets followed a trend almost similar to the volume shrinkage. The linear shrinkage increased from 11% to 26% as temperature increased from 1000°C to 1400°C. It is to be noted that the molar volume of α-Al_2O_3, rutile-TiO_2 and Al_2TiO_5 are 25.56, 28.82 and 49.20 cm³/mol, respectively, which result in a net volume expansion, on transformation to Al_2TiO_5. It is clear from the XRD patterns that the phase transformation to Al_2TiO_5 phase occurs in the range of 1000°C to 1400°C. The linear shrinkage in this temperature range shows this effect, as it decreases from 25% at 1200°C to 22% at 1300°C, and then again increases to 32% at 1500°C.

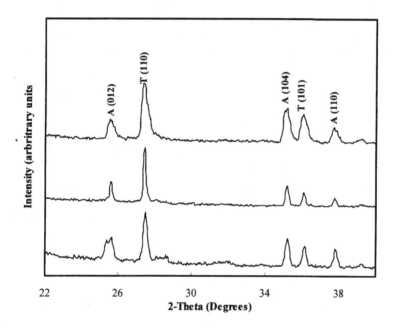

Figure 6a. XRD patterns of the nanocomposite sintered at 1000°C, 1100°C and 1200°C for 2 h. A and T represent α-Al_2O_3 and rutile-TiO_2, respectively.

The pellets used for sintering and densification study were ground into a fine powder, to study the effects of sintering temperature on phase transformation using XRD technique. XRD analysis was also performed to determine the crystallite size. Figure 6a shows the XRD patterns of samples sintered at 1000°C, 1100°C and 1200°C for 2 h. The peaks in the XRD patterns are distinct and sharp indicating complete crystallization and all the phases have been identified. The phases present in the sintered structures were α-Al$_2$O$_3$ and rutile-TiO$_2$. XRD plots of powder heat treated at 1300°C, 1400°C and 1500°C for 2 h, is shown in Figure 6b. Aluminum titanate phase begins to appear at 1300°C and as temperature increase its percentage increases. Initially at 1300°C, percentage of aluminum titanate phase was 15%, which sharply increased to 65% at 1400°C and at 1500°C, pure aluminum titanate phase was obtained. The molar volume of α-Al$_2$O$_3$, rutile-TiO$_2$ and Al$_2$TiO$_5$ are 25.56, 28.82 and 49.20 cm^3/mol resulting in volume expansion on transformation to Al$_2$TiO$_5$ at elevated temperatures.[5] This volume expansion leads to development of compressive stresses and associated strain, resulting in peak shift to higher angles, as observed in figure 6b between 1300°C and 1400°C.

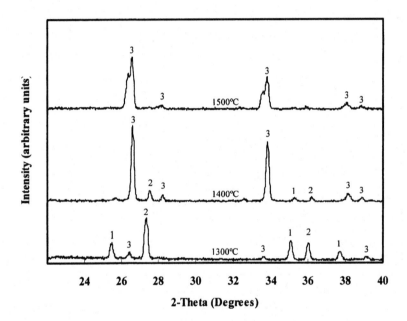

Figure 6b. XRD patterns of the Al$_2$O$_3$-TiO$_2$ nanocomposite sintered at 1300°C, 1400°C and 1500°C for 2 h. 1, 2 and 3 represent α- Al$_2$O$_3$, rutile-TiO$_2$ and Al$_2$TiO$_5$ phases, respectively.

The crystallite size of the powder calcined at various temperatures for 2 h, was calculated using the Scherrer's formula (2)

$$B_{crys} = (k \, \lambda) \, / \, (L \, \cos\theta) \qquad (2)$$

where λ is the wavelength of the X-ray used (1.5406 Å), θ is the Bragg angle, k is a constant and L is the full width at half maximum. The crystallite size at 700°C of anatase-TiO_2 calculated using (101) peak as the reference, was 18 nm and of θ-Al_2O_3 calculated using (102) peak as reference was 26 nm. With the increase in sintering temperature, phase transformation occurred in both α-Al_2O_3 and rutile-TiO_2 resulting in particle size of 24 nm at 1200°C, for both phases. Further increase in the sintering temperature resulted in grain coarsening of the two phases and evolution of aluminum titanate phase. In the specimens sintered at 1300°C, the grain size of α-Al_2O_3, rutile-TiO_2 and AT (calculated using (110) peak as reference) phases were found to be 48 nm, 45 nm and 54 nm, respectively. The results clearly indicate that nano-features were retained in the composite form even after pressure-less sintering for 2 h, at 1300°C.

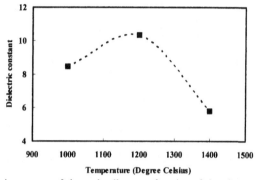

Figure 7. Dielectric constant of sintered pellets as a function of sintering temperature.

Dielectric constants of the sintered pellet structures have peak value at 1200°C (Figure 7) with the highest value of 10.32, in pellets sintered for 2 h. Dielectric constants of bulk materials depend on phase composition, defects, density and grain size.[10] The phases present at both 1000°C and 1200°C are crystalline α-Al_2O_3 and rutile-TiO_2 but the pellets have a marked difference of density, with the density at 1200°C higher. Even though the density increases at 1400°C, the dielectric constant shows a sharp decrease. This is clearly because of the increased volume fraction of the Al_2TiO_5 phase. It must be pointed out that pellets sintered for 2 h have a composite composition of α-Al_2O_3, rutile-TiO_2 and Al_2TiO_5. This seems to be a result of higher density and the affect of volume expansion, due to the phase transformation taking place. The dissipation factors of the pellets sintered at 1200°C and 1400°C, were found to be 0.52 and 0.02, respectively. The formation of Al_2TiO_5 phase has a favorable effect on the dissipation factor and it decreases sharply. Thus, though the formation of Al_2TiO_5 phase decreases the dielectric constant, it improves the dissipation factor. Thus, a proper phase composition can be identified to maintain a balance between the two, according to the requirements.

CONCLUSIONS

Al_2O_3-TiO_2 nanocomposite powders were successfully synthesized via a simple sol-gel process. The effect of pH on the synthesis route has been investigated. Best results were obtained for powders synthesized from sol maintained at a pH of 5.5. Al_2TiO_5 phase began to appear at 1300°C and its phase percentage increased with the increasing sintering temperature, resulting in pure Al_2TiO_5 at 1500°C. The densification study showed the highest density of 2.99 g/cm^3 in pellets sintered at 1500°C, for 2 h. Sintered density increased with the increase in sintering temperature from 1000°C to 1500°C. Results also showed that the presence of Al_2TiO_5 decreases the dielectric constant but significantly improved the dissipation factor. Highest dielectric constant of 10.32 was achieved in pellets sintered at 1200°C, for 2 h. Crystallite-size of Al_2O_3, TiO_2 and Al_2TiO_5 in structures sintered at 1300°C was calculated to be 48 nm, 45 nm and 54 nm, respectively, using the Scherrer's formula.

REFERENCES

[1]P. Singer,"High K dielectrics: No easy solution", *Semiconductor International*, 38 (2003)

[2]R. W. Murto, M. I. Gardner, G. A. Brown, P. M. Zeitzoff, H. R. Huff, "Challenges in gate stack engineering", *Solid state Technology*, 43-48 (2003)

[3]M. Wallace, "Challenges for the characterization and integration of high-k dielectrics", *App. Surface Sc.*, 231–232, 543–551(2004)

[4]L. Stanciu, J. R. Groza, L. Stoica , C. Plapcianu, "Influence of powder precursors on reaction sintering of Al_2TiO_5", *Scripta Materialia*, 50, 1259–1262 (2004)

[5]R. G. Duan, G. D. Zhan, J. D. Kuntz , B. H. Kear, A. K. Mukherjee, "Processing and microstructure of high-pressure consolidated ceramic nanocomposites", *Scripta Materialia*, 51, 1135–1139 (2004)

[6]D. M. Ibrahim, A. A. Mostafa, T. Khalil, "Preparation of tialite (aluminium titanate) via the urea formaldehyde polymeric route", *Ceramics International,* 25, 697-704 (1999)

[7]T. S. Liu, D.S. Perara, "Long-term thermal stability and mechanical properties of Aluminium titanate at 1000-1200°C", *J. Mat. Sc.*, 33, 995-1001 (1998)

[8]C. Suryanarayana and C. C. Koch, *Hyperfine Interactions*, 130, 5 (2000)

[9]J. G. Li and X. Sun, "Synthesis and sintering behavior of a nanocrystalline α-alumina powder", *Acta mater*, 48, 3103-3112 (2000)

[10]C. Hube, C. Elissalde, V. Hornebec, S. Mornet, M. Treguer-Delapierre , F. Weill, M. Maglione, "Nano-ferroelectric based core–shell particles: towards tuning of dielectric properties," *Ceramics International*, 30, 1241–1245 (2004)

SYNTHESIS AND CHARACTERIZATION OF NANOCRYSTALLINE BARIUM STRONTIUM TITANATE CERAMICS

Vikas Somani and Samar J. Kalita

Department of Mechanical, Materials and Aerospace Engineering

University of Central Florida

Orlando, FL 32816-2450

ABSTRACT

Nano-grained Barium Titanate based ceramics are of interest, for applications in ultra thin dielectric layers. In this work, we have synthesized nanocrystalline Barium Strontium Titanate ($Ba_{0.7}Sr_{0.3}TiO_3$) powder in the range of 15 - 25 nm using a simple sol-gel processing route. Barium acetate, strontium acetate and titanium isopropoxide were used as precursors. By varying the pH of the sol and the calcination temperature a simple sol-gel synthesis was developed which can be easily repeated. Thermal properties of the processed gel were determined using Differential Scanning Calorimetry and Thermogravimetric analysis. The $Ba_{0.7}Sr_{0.3}TiO_3$ gel obtained was dried at $200^{\circ}C$, to form powder and subsequently calcined in the temperature range of $400^{\circ}C$ to $700^{\circ}C$ for crystallization. X-Ray Diffraction technique was used to study the phase evolution and phase purity during synthesis. Crystallite size of the powder was also determined using X-ray diffraction patterns. XRD patterns showed variation in phase evolution as a function of pH of the solution. Scanning Electron Microcopy of the synthesized powder calcined at $700^{\circ}C$ showed that the powder was in agglomerates, which were consisted of very fine particles. High Resolution Transmission Electron Microscopy (HR-TEM) results showed that the particle size of the $Ba_{0.7}Sr_{0.3}TiO_3$ powder obtained after calcination at $700^{\circ}C$ was in the range 15-25 nm.

INTRODUCTION

Ferroelectric Barium titanate based dielectrics show tunable dielectric properties and low dielectric losses at room temperature. They find potential applications in capacitors, phase shifters and resonators. Barium strontium titanate (BST) is a solid solution of barium titanate and strontium titanate. This material is lead-free and hence, environmental-friendly.

The hurdle in the size reduction of dielectric resonators is to find materials with high dielectric constant but low dielectric loss. At low temperatures the dielectric loss in polycrystalline ceramics is controlled by extrinsic factors like defects, impurities and grain boundaries. However, Alford et al. showed that the dielectric loss in polycrystalline alumina could be improved by carefully controlling the processing procedure and purity.[1]

Various processes have been used to synthesize BST ceramics and thin films like solid state techniques, hydrothermal process, sol-gel process, sputtering and vapor deposition processes. Sol-gel technique has the advantage of low temperature synthesis, ease in doping, control of structure and potential application in the formation of thin films. Moreover sol-gel is capable of nanosize particles with tailored chemistry and surface properties. The critical issues for wider acceptance of sol-gel science are the cost-factor, control of the processing variables which offers further control on its particle size and properties. Some of the variables that need to be optimized and effectively monitored during the processing are temperature, pH, concentration and solvents used.

The temperature, pH, precursors, additives, mechanical agitation and solvents used during the synthesis of powders have important affects on the properties of the powders produced. Hayashi et al. successfully synthesized 30 - 40 nm size $Ba_{0.7}Sr_{0.3}TiO_3$ powders by sol gel process.[2] $Ba(OH)_2$ and $Sr(OH)_2$ were used for preparing precursor solution. Yang et al. deposited crystalline BST thin films of 200 nm thickness using sol-gel technique.[3] pH was maintained in the range of 2-4 during the process. Selvam et al. synthesized nanopowders of BST at low temperatures of around 100°C. The solvents used in the process were acetic acid and isopropyl alcohol.[4] In another report, $BaNO_3$, $SrNO_3$ and tetrabutyl titanate were dissolved in citrate acid to prepare the precursors. The precursors were allowed to react with the pH of the solution maintained at 6. The resulting solution was dried and then calcined at 800°C to obtain pure $Ba_{0.5}Sr_{0.5}TiO_3$ phase.[5]

Though synthesis of BST ceramics and thin films have received ample attention,[6-9] systematic studies investigating the effect of pH variation on the phase ($Ba_{0.5}Sr_{0.5}TiO_3$) evolution and synthesis of nanocrystalline BST, using barium acetate and strontium acetate as precursor solutions, is limited. Additionally, though researchers have synthesized nanocrystalline BST, synthesis of nano powder in the lower end of nano regime (< 30 nm), is still a challenge. Here, we report a simple method to synthesize $Ba_{0.7}Sr_{0.3}TiO_3$ nanopowder using sol-gel technique. The effect of pH variation on phase evolution and crystallanity has been investigated. The nanopowder has been characterized for phase purity and crystallinity using XRD. Powder morphology has been analysed using SEM. HR-TEM analysis was employed to analyze the powder morphology and lattice fringes.

EXPERIMENTAL PROCEDURE

Powder Synthesis

$Ba_{0.7}Sr_{0.3}TiO_3$ nanopowders were synthesized using barium acetate (reagent grade, Acros Organics, USA), strontium acetate (reagent grade, Acros Organics, USA) and titanium (IV) isopropoxide (purity 98 +%, Acros Organics, USA) as the starting chemicals via sol-gel processing. Acetic acid (5% v/v Fisher scientific, USA) was used as solvent and 2-methoxy ethanol (99 +%, across Organics, USA) was used as a stabilizer for titanium (IV) isopropoxide solution. Stoichiometric proportions of barium acetate (8 mol/L) and strontium acetate (6.8 mol/L) powders were dissolved separately in acetic acid, by continuous magnetic stirring at 300 rpm for 15 min and at a temperature of 50°C. A magnetic stirrer with a hot plate was used for this purpose. The two solutions were then mixed together and then refluxed at 125 °C for 2 h. 2-Methoxy ethanol (2-4 ml) was added in titanium (IV) isopropoxide (0.011 mol) to form a separate solution, at room temperature. The Ba–Sr solution was added drop-wise into the Ti solution with the help of a burette. The pH was varied in the range of 1 to 7 by adding hydrochloric acid or ammonium hydroxide to the solution. The solution was magnetically stirred at 300 rpm for 15 min, for homogenization. Refluxing of the resultant solution led to the formation of a thick white-colored gel. At this stage, deionized water was added to the gel and the solution was again stirred magnetically for another 20-45 min. The final solution obtained was filtered and then heated to 200°C for 2 h.

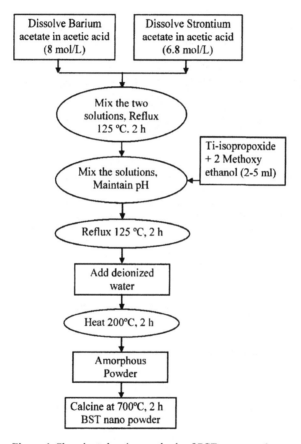

Figure 1. Flowchart showing synthesis of BST nano powder.

This resulted in the formation of amorphous $Ba_{0.7}Sr_{0.3}TiO_3$ powders, later found to be in the nano range. The amorphous powders were calcined at 700°C in a muffle furnace, under atmospheric conditions to form crystalline nanopowders. Figure 1 shows a schematic of the process used to form BST powders by this sol-gel route.

Differential Scanning Calorimetry and Thermogravimetric analysis

As-formed BST gel was characterized using Differential Scanning Calorimetry (DSC) and Thermogravimetric Analysis (TGA) to ascertain the thermal stability and decomposition temperature of various species. DSC is a thermal analysis technique that measures the energy absorbed or emitted by a sample, as a function of temperature or time. When thermal transition occurs in the sample, DSC provides a direct calorimetric measurement of the transition energy at the temperature of the transition. DSC thermal analysis is very helpful to determine calcination temperature and time ideal for evolution of phases. In order to determine material specific calcination temperature, to form crystalline $Ba_{0.7}Sr_{0.3}TiO_3$ powders, we analyzed the thermal properties of the BST gel from room temperature to 750°C. For this, as-processed gel sample was placed in the specimen holder of a DSC/TGA analyzer (Model SDT Q600 from TA Instruments, Inc.) and heated from room temperature to 750°C, at a heating rate of 6°C/min. Argon was used for purging at 10 ml/min.

Powder Characterization

Characterization of the synthesized powder was done using X-Ray Diffraction (XRD), Scanning Electron Microscopy (SEM) and High-Resolution Transmission Electron Microscopy (HR-TEM) techniques. As-produced $Ba_{0.7}Sr_{0.3}TiO_3$ powder from sol prepared at different pH was studied for phase evolution and phase purity, using X-ray powder diffraction. For this, agglomerates of the synthesized powder produced from sol maintained at pH of 1.0, 3.5, 5.5 and 7.0 and calcined at 700°C, were ground into fine powder using a mortar and pestle. Fine powders obtained were analyzed with a Rigaku diffractometer (Model D/MAX-B, Rigaku Co., Tokyo, Japan) using Ni filtered CuKα radiation (λ =0.1542 nm) at settings of 40 kV and 40 mA, supported by JADE software. The XRD patterns were recorded in the 2θ range of 20-40 degrees and were used for phase analysis. The crystallite-size of the powder was calculated employing the peak broadening, using JADE software and the Scherrer's formula.

Based on XRD results, powder obtained from sol with pH of 1.0, and calcined at 700°C for 2 h, was studied for its morphology and crystallite-size, using HR-TEM. Powder was examined using HR-TEM from Phillips (Model Tecnai – Philips F30, FEI Co., Hillsboro, OR).

The morphology of the $Ba_{0.7}Sr_{0.3}TiO_3$ powder, calcined at 700°C for 2 h, was also observed using SEM. The sample powders were gold-palladium coated using a sputter coater at a current of 20 mA for 6 min. The resulting film had a thickness of 60 nm. The powder surfaces were then imaged using a Hitachi S 3500 A SEM.

RESULTS AND DISCUSSION

The pH changes occurring during the reaction were monitored. The initial Ba-Sr-Ti solution was weakly acidic and its pH was in the range of 4.5-5.5. The pH of the solution results from the dissociation and dissolution of precursor ions. The pH of the solution was varied by addition of hydrochloric acid or ammonium hydroxide as required.

The TGA and DSC plots of the as-synthesized BST specimen (pH 1) heated up to 750°C, are shown in Figure 2. The first endothermic peak on the DSC curve was observed between 50°C to 190°C. This region corresponds to a weight loss of about 65% on the TGA curve and is a result of the loss of absorbed water. The second exothermic peak observed from 320°C to 400°C corresponds to a weight loss of about 5% and results from the dissociation of organic compounds and solvents. A third continuous endothermic heat flow region begins from 440°C and ends at around 550°C. In this region, the initial formation of $Ba_{0.7}Sr_{0.3}TiO_3$ phase occurs and is followed by its crystallization. The rate of weight loss after 550°C is very low and is mainly due to the slow elimination of carbonate groups, as detected by XRD. The TGA / DSC analysis show that the as-synthesized powder needs to be sintered above 400°C to get carbon free particles. Based on DSC / TGA analysis, we concluded that a calcination temperature of 400°C would be ideal for the removal of all organics and evolution of the desired $Ba_{0.7}Sr_{0.3}TiO_3$ phase and crystallization.

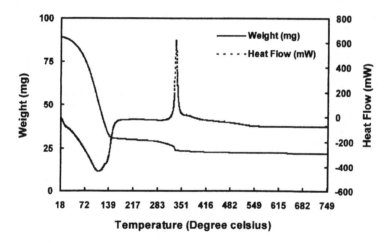

Figure 2. DSC and TGA graphs of as-prepared BST gel.

X-ray powder diffraction analysis was conducted to evaluate the phase evolution and crystallanity of powders during synthesis. Effect of pH change of the sol on evolution of BST phases, was studied. Figure 3 shows the XRD patterns of the powders, synthesized from sol with different pH and calcined at 700°C for 2 h. Peaks in each X-ray diffraction pattern were recorded and verified using standard PDF file. The powder synthesized at pH 7 and 5.5 and calcined at 700°C for 2 h did not show any peak of desired $Ba_{0.7}Sr_{0.3}TiO_3$ phase. The powders are amorphous and there is no sign of formation of crystalline $Ba_{0.7}Sr_{0.3}TiO_3$. As the pH is further decreased to 3.5 by adding hydrochloric acid, the powders show partial crystallinity; i.e. most of the powder is still amorphous. The peaks observed are very broad and these broad peaks of different phases overlap each other. Moreover, the amount of barium carbonate in this synthesis route (pH 3.5) is substantially high.

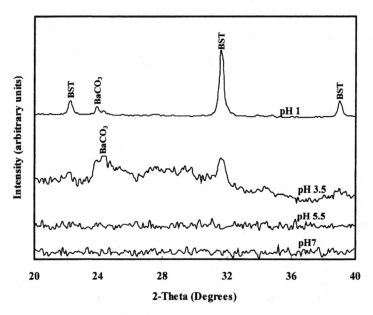

Figure 3. XRD patterns of BST ceramics synthesized at various pH values and calcined at 700°C for 2 h.

Powder, synthesized from the solution maintained at pH 1, show peaks of $Ba_{0.7}Sr_{0.3}TiO_3$. The peaks are broad indicating that the powder is of a fine size. Small amount of barium carbonate was also detected. It is apparent that low pH value enhances the formation of $Ba_{0.7}Sr_{0.3}TiO_3$ crystalline nanopowder. It is known that pH along with mechanical agitation and temperature can influence the kinetics, growth reactions, hydrolysis and condensation reactions during sol-gel synthesis. pH in particular has a major effect on the hydrolysis and condensation reactions. Acidic conditions are known to facilitate hydrolysis and basic conditions favor condensation. Hence, low pH (highly acidic conditions) results in fully hydrolyzed species before condensation begins. In this case, low pH leads to complete hydrolysis of barium and strontium precursors and forbids their condensation into respective oxides or carbonates. Proper mechanical agitation and temperature, along with the fact that sol-gel technique allows molecular mixing, results in the synthesis of the desired phase; instead of carbonates, as seen during synthesis with pH at 3.5. It is to be noted that the only parameter altered during the synthesis route was pH. Other parameters like concentration, temperature, mechanical agitation and heat treatment were the same. All these variables can significantly affect the synthesis process. Effect of varying any or all of these parameters while maintaining the same pH, can result in varied results.

The crystallite size of the powder was calculated using the Scherrer's formula using the equation (1).

$$B_{crystal} = (k\,\lambda) / (L\,\cos\theta) \qquad (1)$$

Where λ is the wavelength of the X-ray used (1.5406 $A°$), θ is the Bragg angle, k is a constant and L is the full width at half maximum. The crystallite size of the powder, considering the full width at half maximum, was calculated to be 25 nm.

SEM micrograph of the powders calcined at 700°C for 2 h, is shown in Figure 4. The calcined powders are hard agglomerates of fine powder particles and could not be broken by manual grinding with a mortar and pestle. Most of the agglomerates are ~15-30 μm in size. However, it can be easily visualized from this micrograph that these agglomerates are composed of very fine particles with dimensions possibly in the lower end of nano range. HR-TEM technique was used to determine the actual crystallite-size of the powder and to verify the results obtained from XRD analysis using Scherrer's formula.

TEM micrograph of powders calcined at 700°C for 2 h, is shown in figure 5. The crystalline nature of the particles is apparent by the lattice fringes seen in the figure. Particle size was calculated by measuring the number of particles, across its length and breadth. The average particle size thus calculated was 20 nm, which matches well with the XRD results.

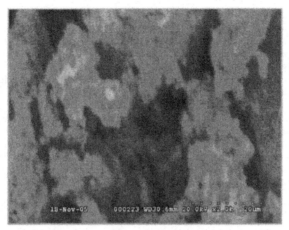

Figure 4. SEM micrograph of the BST powder synthesized at pH 1 and calcined at 700°C for 2 h.

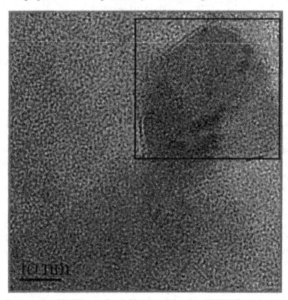

Figure 5. TEM micrograph of BST powder (pH=1) calcined at 700°C for 2 h. Particle size of the powder was found to be 15-25 nm.

CONCLUSIONS

Nanocrystalline $Ba_{0.7}Sr_{0.3}TiO_3$ powder of average size 15-25 nm was synthesized via sol gel processing technique using barium acetate, strontium acetate and titanium isopropoxide as precursors. Processing parameters were optimized and the phase evolution during synthesis was studied. The DSC / TGA analysis showed that the as-processed gel needs to be sintered above 400°C to obtain carbon free powder. Complete crystallization was achieved at 700°C where the phase composition is predominantly $Ba_{0.7}Sr_{0.3}TiO_3$. Phase evolution, as observed in the XRD patterns, were consistent with our findings from DSC / TGA analysis of the BST gel. Based on XRD pattern (considering the full width at half maximum) the crystallite size of the powder was calculated to be ~ 25 nm. SEM analysis showed that the synthesized BST powder were in agglomerates of 15-30 μm in size consisting of very fine powder particles in the nano regime. HR-TEM analysis confirmed that the particle (crystallite) size of the synthesized $Ba_{0.7}Sr_{0.3}TiO_3$ powder, obtained after calcination at 700°C for 2 h were in the range 15-25 nm which is in line with the results obtained using the Scherrer's formula on the X-ray diffraction pattern.

ACKNOWLEDGEMENT

We would like to offer our thanks to Prof. Helge Heinrich of Advanced Materials Processing and Analysis Center (AMPAC) and the Department of Physics for his help with transmission electron microscopy. We also acknowledge experimental support by Ankush Halbe and Narayana Garimella and, editorial help by Chayanika D. Kalita.

REFERENCE

[1]N. Alford and Stuart J. Penn, "Sintered alumina with low dielectric loss", *Journal of Applied Physics*, **80**(10), 5895-5898 (1996).

[2]T. Hayashi, H. Shinozaki and K. Sasaki, "Preparation and Properties of $(Ba_{0.7}Sr_{0.3})TiO_3$ Powders and Thin Films Using Precursor Solutions Formed from Alkoxide-hydroxide", *Journal of the European Ceramic Society*, **19**, 1011-1016 (1999).

[3]W. Yang, A. Chang and B. Yang, "Preparation of Barium Strontium Titanate Ceramic by Sol-Gel Method and Microwave Sintering", *Journal of Materials Synthesis and Processing*, **10**(6), 303-309 (2002).

[4]I. P. Selvam and V. Kumar, "Synthesis of nanopowders of $(Ba_{1-x}Sr_x)TiO_3$", *Materials Letters*, **56**, 1089-1092 (2002).

[5]C. Fu., C. Yang, H. Chen, Y. Wang and L. Hu, "Microstructure and dielectric properties of $BaxSr1.xTiO3$ ceramics", *Materials Science and Engineering B*, **119**, 185–188 (2005).

[6]R. H. Liang, X. L. Dong, Y. Chen, F. Cao, Y. L. Wang, "Effect of various dopants on the tunable and dielectric properties of $Ba0.6Sr0.4TiO3$ ceramics", *Ceramics International*, **31**, 1097–1101 (2005).

[7]H. V. Alexandru, C. Berbecaru, A. Ioachim, M. I. Toacsen, M. G. Banciu, L. Nedelcu and D. Ghetu, "Oxides ferroelectric (Ba, Sr)TiO3 for microwave devices", *Materials Science and Engineering B*, **109,** 152–159 (2004).

[8]C. Shen, Q. F. Liu and Q. Liu, "Sol–gel synthesis and spark plasma sintering of $Ba_{0.5}Sr_{0.5}TiO_3$", *Materials Letters*, **58,** 2302-2305 (2004).

[9]X. Wei, A. L. Vasiliev and N. P. Padturea, "Nanotubes patterned thin films of barium-strontium titanate", *J. Mater. Res.*, **20**(8), 2140-2147 (2005).

NANOPARTICLE HYDROXYAPATITE CRYSTALLIZATION CONTROL BY USING POLYELECTROLYTES

Mualla Öner, Özlem Dogan
Yildiz Technical University
Chemical Engineering Department
Davutpasa, Istanbul, Turkey 34210

ABSTRACT

The synthesis of advanced inorganic materials such as biomedical devices, catalysis, ceramics, optical and electronic films require crystallization strategies that provide control over the size, structure, orientation and morphology of the forming crystal. In biomineralization, inorganic crystals form under the full control of structure directing polymers, such as proteins and polysaccharides. In this work we present a facile way to produce HAP nanoparticles by wet chemical synthesis in the presence of polyelectrolytes under strictly controlled temperature, pH, and atmospheric conditions. The constant-composition method has been used to study the influence of polyelectrolytes on the kinetics of crystal growth of hydroxyapatite (HAP) on HAP seed crystals at pH 7.4 and 37 °C. The results indicate that polyelectrolyte concentration and the larger number of negatively charged functional groups markedly affect the growth rate. The fit of the Langmuir adsorption model to the experimental data supports a mechanism of inhibition through molecular adsorption of polymers on the surface of growing crystals.

INTRODUCTION

Hydroxyapatite ($Ca_5(PO_4)_3OH$, HAP) is a compound of great interest because it is the main inorganic constituent of human bones, teeth and soft tissues[1]. As a compound with structural and chemical resemblance of bone mineral, HAP is of particular importance in the field of biomaterials. The properties that make HAP superior as a biomaterial in contrast to the metals or bioinert ceramics are absence of toxicity, biocompability with hard and soft tissues and bioactivity[2]. It directly bonds to the bone and favours the implant fixation. Non-medical applications include packing media for protein column chromatography, gas sensors, water treatment, catalysts and ion exchangers due to its unique surface structure and ionic substitutions[3]. The growth mechanism of HAP has received considerable attention in view of its importance in understanding the mechanism of hard tissue calcification and in many undesirable cases of pathological mineralization of articular cartilage, formation of dental caries, formation of renal, bladder and bile stones, atheromatic plaque and calcification of transplanted cardiac valves[4].

The molecular control of inorganic crystallization by organic substances is a key technology for the fabrication of novel materials that has recently received a considerable amount of attention[5]. This process mimics biological mineralization in which a preorganized organic phase provides a niche for inorganic crystals to nucleate and grow. Many of the acidic macromolecules isolated from biomineral have demonstrated ability to regulate mineral growth in vitro[6]. The size and shape of the crystals are thought to be controlled by the size and shape of the organic matrix which acts as a template for the mineral microstucture[7]. Nucleation may be initiated by the adsorption of cations onto functional sites of acidic macromolecules which promotes the formation of critical nuclei[8]. Crystallite orientation may be controlled by specific molecular interactions which causes the arrangement of solution ions in specific locations relative to organic sites. In the presence of biological macromolecules such as collagen, the nucleation and growth of nanocrystalline apatite to form highly organized bone minerals is one of the most fascinating process in nature[9].

One problem in nucleation and crystal growth, of considerable interest in biological situations, concerns the way in which certain polyelectrolytes affect crystallization from supersaturated solutions of inorganic salts[10-11]. As many biological molecules contain oxyanion functional groups (carboxylate, phosphate and sulphate esters) we have been involved in crystallization experiments using polyelectrolytes.

The main aim of this work was to control nucleation and growth of HAP in the presence of dissolved synthetic polyelectrolytes mimicking the action of natural proteins, under strictly controlled temperature, pH, and atmospheric conditions.

EXPERIMENTAL METHODS

Preparation of polymers

Acidic acrylate block copolymers have been made, by radical polymerization, with defined molecular weight and structure. Radical polymerization of acrylic acid (AA) was carried out in the presence of α-thio polyethylene glycol monomethylether as a chain transfer agent to produce Poly(ethylene glycol -block -acrylic acid) copolymers. PEG block length in the copolymers was controlled by using three different molecular weight chain transfer agents (M_n:350, 750 and 2000 g/mole). The complete experimental procedures were reported previously[12].

Preparation of HAP seed crystals

HAP seed crystals were prepared from calcium chloride and dipotassium hydrogen phosphate solution. 100 ml of 0.5 M $CaCl_2$ and 0.1 M KOH to adjust pH at 9.5 were added into reaction vessel, which was thermostated at 70 °C. This solution was mixed with 0.3 M dipotassium hydrogen phosphate solution, added dropwise over a period of 2 hours. During the precipitation process, the pH was kept constant between 9.0-9.5 by small additions of KOH solution. During precipitation, presaturated nitrogen gas was introduced into the stirred solution to ensure a CO_2 free atmosphere. In experiments where different polymers were used for obtaining hydroxyapatite seeds a similar procedure was followed and freshly prepared polymer solutions were normally added to the phosphate solution at concentration of 4 ppm. The precipitate was separated and refluxed for twenty four hours in its supersaturated solution, aged for one month at 37 °C, while pH was kept at about 7.0. The crystalline solids were analyzed by a number of techniques. X-ray diffraction analysis was carried out using Cu-Kα radiation in a Phillips Panalytical X'ert Pro powder diffractometer operating at 40 mA and 40 kV. The 2θ range was from 20° to 60 ° at scan rate of 0.020° step^{-1}. Purity of the samples was also tested FTIR spectral analysis. One milligram of powdered samples was carefully mixed with 100 mg of KBr of (Merck, infrared grade) and pelletised under a pressure. The pellets were analyzed using a Perkin Elmer Spectrum One in the 4000-400 cm^{-1} region at a resolution of 4 cm^{-1} .The specific surface area (SSA) of the synthesized samples was determined by nitrogen sorption/desorption according to multiple point BET method (Quantachrome Instruments Autosorb 1). The crystals were outgassed for 24 h at 40 °C. The seed crystals were characterized by scanning electron microscopy (JEOL-FEG-SEM),

The kinetics of precipitation of HAP on seed crystals

Constant-composition crystal growth experiment was used for determination of the growth rate on HAP seed crystals. In a typical experiment, stable supersaturated solutions of calcium phosphate with a molar ratio $Ca_t/P_t=1.67$ were prepared in water-jacketed reactor. The total molar concentration of calcium, Ca_t was 5.00×10^{-4} mole/liter with calcium/phosphate molar ratio 1.67. Following the pH adjustment and verification of the stability of the supersaturated solution for at least 2 h, known weights of HAP seed crystals was added to the solution. As growth commenced, the release of protons lowered the pH of the solutions and a pH drop triggered the addition of titrant solutions, which was controlled by means of pH-stat (Radiometer pH-Stat Titrator PHM290, Auto-Burette ABU901). The crystal growth reaction was monitored by the addition of titrant solutions as a function of time from mechanically coupled automatic burettes. The rates of crystallization were determined from the rates of addition of mixed titrants and corrected for surface area changes. The complete experimental procedures were reported previously[13].

RESULTS AND DISCUSSION
XRD Results

XRD spectra of all samples have shown peaks characteristic for HAP. The grown needlelike crystals were identified as Hydroxyapatite by XRD (Fig.1) and compared with that of the ASTM (09-0432) standards.

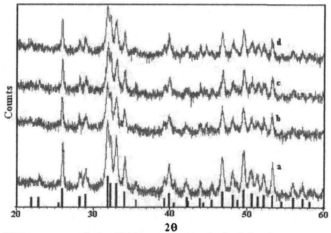

Figure 1. XRD patterns of the HAP samples synthesized in the presence of different polyelectrolytes (a) Crystals grown from a solution, HAP (b) Crystals grown from a solution containing 4 ppm poly(ethylene glycol-b-acrylic acid) copolymer, HAPEO-b-AA (c) Crystals grown from a solution containing 4 ppm poly(acrylic acid) homopolymer, HAPPAA (d) Crystals grown from a solution containing 4 ppm poly(methacrylic acid) homopolymer, HAPPMAA.

All intensity peaks of the XRD patterns of the HAP powders produced were exactly matched with the structural data of the HAP described in ASTM standards, although the intensities of the XRD peaks were varied with the reaction conditions. Lattice parameters of the crystals precipitated are a=b=9.4172 Å and c=6.8799 Å. These values very close to the theoretical values (a=b=9.4180 Å and c= 6.8840 Å). The theoretical values for HAP are from ASTM card file No. 9-432.

The mean crystallite sizes were calculated using well-known Scherrer Equation[14]

$$L = \frac{k\lambda}{BCos\theta} \tag{1}$$

where L is the mean crystallite size in nm, k is the shape factor, B is the broadening of the diffraction line measured at half of its maximum intensity. B was determined by full width at half maximum (FWHM) middle point method. The shape factor k becomes 0.9 when FWHM is used for B. λ is the wavelength of X-ray and θ is the Bragg diffraction angle[15].

The size of the individual HAP crystals was calculated from the broadening 002 and 300 diffraction peaks, assuming that HAP crystals were hexagonal prisms with height equal to crystallite size along 002 planes (c-axis) and width of prism corresponding to crystallite size along 300 planes (a-axis) (Table I). The results showed that L_{002} values of HAP sample is significantly reduced in the presence of polymers. The reduction in size is greater in the direction of the c-axis. L_{300} values of samples do not change in the presence of polymer except the sample synthesized using PAA.

Table I. The relative intensity (I/II) and the crystallite size (L) evaluated from the width at half maximum intensity (B) of the (002) and (003) reflection of HAP synthesized using different polyelectrolytes

Polymer	I/II	L_{002} (nm)	$1/B_{002}$	I/II	L_{300} (nm)	$1/B_{300}$	L_{002}/ L_{300}
HAPBlank	57	83	10.16	65	30	3.63	2.77
HAPPAA[a]	56	59	7.26	54	53	6.35	1.11
HAPPMAA[b]	57	52	6.35	56	26	3.18	2.00
HAPEO-b-AA[c]	55	52	6.35	56	30	3.63	1.73

a: PAA(Polyacrylic acid, M_n=5000); b:PMAA(Polymethacrylic acid, M_n=8000);
c:EO-b-AA, Poly(ethylene glycol -block -acrylic acid, M_n=4200, EO/AA=0.35)

The crystallinity of hydroxyapatite synthesized at different conditions is influenced by the crystallite shape and size. Therefore crystallinity can be expressed as the reciprocal of the line broadening, B^{-1}, at the (002) and (300) diffraction peaks. The crystallinity decreases in the presence of the polymers as can be seen in Table I and X-ray figures. The crystallite growth was anisotropic in space. The size of the crystallites responsible for the Bragg reflection (002) and (300) planes were 83 nm and 30 nm respectively in the absence of polymers. The (c) axis direction was about 2.77 times longer than (a) axis direction (L_{002}/L_{300}) and anisotropy decreases in the presence of polymers (Table I).

The relative intensity means the relative peak intensity ratio measured by comparing the intensity at the (002) or (300), I with that at the (211) face, II. The standard value of the relative intensity at each specific face of HAP represents 0.4 for (002), 0.6 for (300) and 1.0 for (211). At the (002) face, the relative intensity of the HAP crystals was always higher than the standard value, regardless of the different polymers used during production of the crystals. However, relative intensity at the (300) face was observed to be lower than the standard value in the presence of the polymer. These results indicated that the crystal growth of HAP was prominent on the (002) face in the direction of the crystallographic c-axis during the crystallization.

FTIR results

FTIR spectra of the seed crystals also showed that they consisted of HAP (Fig.2). In general, the FTIR spectra do not show differences on absorption bands due to different polymers used. The bands at 3570 cm^{-1} are characteristics for stretching and vibration modes of the OH$^-$ ions. Characteristic frequencies derived from PO_4^{3-} modes can be seen at 1094 cm^{-1}, 1035 cm^{-1}, 962 cm^{-1}, between 600-560 cm^{-1}. The IR peak at 875 cm^{-1} characteristic for HPO_4^{2-} was observed for the crystal. CO_3^{2-} ions were detected in the precipitate from the peaks at 1453 cm^{-1}. [16]

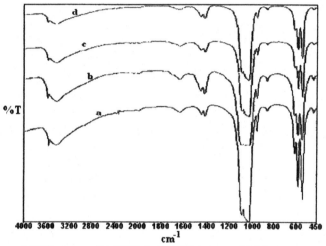

Figure 2. FTIR spectra of (a)HAP (b) HAPEO-b-AA (c) HAPPAA (d) HAPPMAA

Morphology and size of the crystals

Polyelectrolytes have shown significant effect on crystal size. Photographs were also taken by scanning electron microscope (SEM) for the subsequent visual analysis in order to asses the effects of polymers on crystal shape and size (Fig.3. The crystals precipitated at 37 °C have the form of tiny needles with average length of around 191 nm and width of 52 nm (Fig 3a). Figures 3 b-d shows the effect of polymers on crystal habit. No apparent changes in morphology were observed in the presence of polyelectrolytes. The precipitated crystals show similar needlike shape regardless of the different polymers used. The average length of the crystals grown in the presence of the polymers was significantly less than that of the control sample. The average length of the crystals was reduced to 184 nm, 133 nm and 156 nm for solution containing 4 ppm PMAA, EO-b-AA and PAA respectively. The average value of the width in the samples containing polymers was similar to that of the control sample. The aspect ratio (height, H)/(width, W) slightly decreases in the presence of polymers.

Influence of polyelectrolytes on HAP growth kinetics

In this work the influence of trace amounts of polymers on the crystal growth of HAP has been studied by the seeded growth technique. The crystal growth experiments using HAP seed crystals were all made at constant ionic strength and constant supersaturation with respect to HAP. Previous studies have been made of the role of serine, tyrosine and hydroxyproline amino acids with polar side groups[17-20]. Results of these kinetic studies suggest that the reduction in the rate of crystal growth is due to the adsorption of amino acids molecules on the surface of HAP seed crystals. Assuming a steady-state adsorbtion/desorbtion and the absence of interaction between growth sites, a Langmuir adsorption isotherm can be tested by plotting $R_0/(R_0-R_i)$ against C_i^{-1}, where R_i and R_0 are the rates of HAP crystallization in the presence and absence of polymers, respectively, and C_i is polymer concentration. The linearity of the resulting plot in Figure 4 suggests that all polymers retard the rate of crystallization by adsorption at active growth sites on the crystal surfaces. The low value of the reaction rate for PAA may reflect stronger equilibrium adsorption of PAA than of PMAA on the HAP surface.

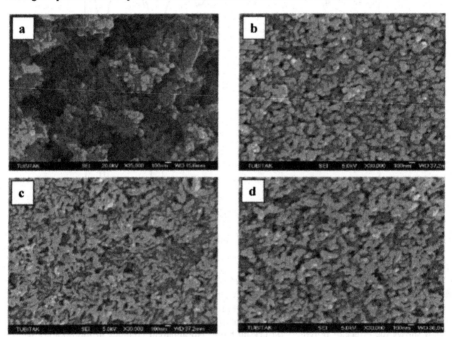

Figure 3. SEM of the HAP samples synthesized in the presence of different polyelectrolytes (a)HAP (b) HAPEO-b-AA (c) HAPPAA (d) HAPPMAA

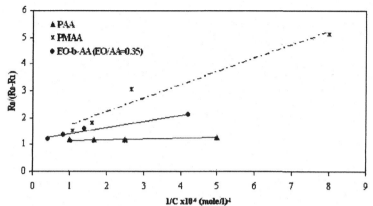

Figure 4 Crystal growth of HAP on HAP seed crystals. Langmuir adsorption isotherm for the effect of polymers.

CONCLUSIONS

Different HAP seed crystals were synthesized using a wet precipitation method in the presence of different polyelectrolytes. The results indicate that polymers are effective in controlling the size and crystallinity of the resulting HAP particles. Kinetic studies suggest that the higher affinity of PAA for HAP corresponds to the more significant effect of this polymer on the rate of HAP crystal growth. From a physicochemical point of view, efficient regulators of calcification would be those molecules that have a high enough adsorption affinity so that low concentrations can result in the adsorption on and blocking of a significant number of crystal growth sites on the crystallites of the tissue.

ACKNOWLEDGEMENTS

We appreciate the support of YTUAF (Project No: 22-07-01-01); TUBITAK (Project No: TBAG-AY/236(101T098); TBAG-AY/250(101T174)) and DPT (Project No: 24-DPT-07-04-01) for the accomplishment of this work. The authors thank Prof.Dr. Sabriye Pişkin of Yildiz Technical University for use of the XRD and FTIR.

REFERENCES

[1]P.G. Koutsoukos, Z. Amjad, M.B. Tomson, and G.H. Nancollas, "Crystallization of Calcium Phosphates. A Constant Composition Study", *J. of the American Chemical Society*, **102(5)**, 1553-1557 (1980).

[2]S. Koutsopoulos, and E. Dalas, "The Crystallization of Hydroxyapatite in the Presence of Lysine", *Journal of Colloid and Interface Science*, **231**, 207-212 (2000).

[3]K. De Groot, C.P.A.T. Klein, J.G.C. Wolke, and J.M.A. De Blieck-Hogervorst, "Chemistry of Calcium Phosphate Bioceramics", *CRC Handbook of Bioactive Ceramics*,**Vol. II**, CRC Pres, Boston (1990).

[4]S. Koutsopoulos, and E. Dalas, "Hydroxyapatite Crystallization in the Presence of Serine, Tyrosine and Hydroxyproline Amino Acids with Polar Side Groups", *Journal of Crystal Growth*,

216, 443-449 (2000).

[5]M. Sedlak, H. Colfen, "Synthesis of Double Hydrophilic Block Copolymers with Hydrophobic Moieties fort the Controlled Crystallization of Minerals", Macromol.Chem.Phys., 202, 587-597 (2001).

[6]S. Weiner, L. Addadi, "Acidic Macromolecules of Mineralized Tissues: The Controllers of Crystal Formation", Trends in Biochemical Sciences,16(7), 252-256 (1991).

[7]B.R. Heywood, S. Mann, "Template-Directed Nucleation and Growth of Inorganic Materials", Advanced Materials, 1, 9-20 (1994).

[8]G. Hunter, "Interfacial Aspects of Biomineralization", Current Opinion in Solid State and Materials Science, 1(3), 430-435 (1996).

[9]P. Calvert,"Biomimetic Ceramics and Composites", MRS Bulletin, 10, 37-40, (1992).

[10]M. Öner, and P. Calvert, " Effect of Architecture of Acrylic Polyelectrolytes on Inhibition of Oxalate Crystallization", Materials Science & Engineering C, Biomimetic Materials, Sensors and Systems, C.2, 93-102, (1994)

[11]M. Öner, J. Norwig, W.H. Meyer, G. Wegner, "Control of ZnO Crystallization by a PEO-b-PMAA Diblock Copolymer", Chem. Mater., 10(2), 460-463, (1998).

[12]E. Akyol, A. Bozkurt, and M. Öner, "The Effects of Polyelectrolytes on the Inhibition and Aggregation of Calcium Crystallization", Polymers for Advanced Technologies, 17, 58-65, 2006

[13]Ö. Doğan, and M. Öner, "Inhibitory Effect of Polyelectrolytes on Crystallization Kinetics of Hydroxyapatite", Progress in Crystal Growth and Characterization of Materials, 50(1-3), 39-51, (2005).

[14]A. Bigi, E. Foresti, M. Gandolfi, M. Gazzano, and N. Roveri, "Inhibiting Effect of Zinc on Hydroxylapatite Crystallization", Journal of Inorganic Biochemistry, 58, 49-58, (1995).

[15]H.P. Klug, L.E. Alexander, X-ray Diffraction Procedures, John Wiley&Sons Inc., New York, (1974)

[16]E. Bertoni, A. B,g,, G. Falini, S. Panzavolta, N.Roveri, "Hydroxyapatite/polyacrylic acid nanocrystals, Journal of Materials Chemistry, 9, 779-782 (1999).

[17]Z. Amjad, "Constant-Composition Study of Dicalcium Phosphate Dihydrate Crystal Growth in the Presence of Poly(acrylic acids)", Langmuir, 5, 1222-1225 (1989).

[18]S. Koutsopoulos, E. Pierri, E. Dalas, N. Tzavellas, and N. Klouras, "Effect Ferricenium Salts on the Crystal Growth of Hydroxyapatite in Aqueous Solution", Journal of Crystal Growth, 218, 353-358 (2000).

[19]S. Koutsopoulos, and E. Dalas, "The Crystallization of Hydroxyapatite in the Presence of Lysine", Journal of Colloid and Interface Science, 231, 207-212 (2000).

[20]S. Koutsopoulos, and E. Dalas, "The Effect of Acidic Amino Acids on Hydroxyapatite Crystallization", Journal of Crystal Growth, 217, 410-415 (2000).

SYNTHESIS OF CARBON NANOTUBES AND SILICON CARBIDE NANOFIBERS AS COMPOSITE REINFORCING MATERIALS

Hao Li
Department of Mechanical and Aerospace Engineering
University of Missouri at Columbia
Columbia, MO, 65211

Abhishek Kothari, and Brian W. Sheldon
Division of Engineering
Brown University
Providence, RI 02906

ABSTRACT

The excellent mechanical properties of nanomaterials are driving research into the creation of strong and tough nanocomposite systems and new forms of nanomaterials. It is critical to select an appropriate reinforcing material with desired microstructures and properties to achieve better nanocomposite performance. The present study focused on synthesis and processing-microstructure relationships of multiwalled carbon nanotubes (CNTs) and SiC nanofibers with chemical vapor deposition (CVD). Various CNTs grown by CVD with anodic aluminum oxide (AAO) templates were examined with scanning and transmission electron microscope (SEM and TEM). It was demonstrated that the experimental conditions, especially catalysts and plasma, have significant impact on CNT growth rates and microstructures. Both catalyst and plasma can increase the deposition rate about one order of magnitude. In addition, catalysts promote the secondary growth of CNTs inside the primary CNTs and plasma may improve the stiffness of primary and secondary CNTs. The SiC nanofibers grown by CVD with catalysts were also investigated. SiC diameters match well with the diameters of precursor CNTs, indicating SiC nanofiber size is controlled by the catalysts originated from CNTs. Generally, CNTs fabricated with CVD-template method have disordered graphitic structures and thus have lower tensile strength, but the disordered structures may help for the load transfer between graphitic layers. In the present study, CVD-SiC nanofibers appear to be stiffer than multiwalled CNTs and may also serve as a good candidate for composite reinforcing materials.

INTRODUCTION

The high modulus and strength of nanomaterials, such as carbon nanotubes (CNTs) and carbon nanofibers (CNFs), has sparked tremendous interest in nanomaterial reinforced composites and development of novel materials for composite applications. Numerous efforts worldwide are addressing all aspects of this rapidly developing field, including synthesis, processing, characterization and integration.[1-2] Reinforcing materials, as one of the three major components of the composite system, strongly affect composite properties, such as strength, toughness, conductivity, and thermal stability. The synthesis and selection of reinforcing materials is thus critical to composite design and manufacturing. CNTs and CNFs are the most widely investigated reinforcing materials for ceramic, metal, and polymer based nanocomposites, while other materials, such as clay, have also been intensively investigated as a reinforcing material.

41

A good understanding of the composite system is required for the design and synthesis of reinforcing nanomaterials considering the objectives of the inclusion of nanomaterials vary for different systems. In a polymeric nanocomposite, the application of CNTs is expected to increase its strength for nanotube super strength and stiffness, while improving or at least keeping its toughness for nanotube flexibility. The extremely large interfacial area distinguishes nanocomposites from traditional microfiber reinforced composites. The impact of the interface over a few R_g, (radius of gyration, in the order of tens of nanometers) in polymer matrix is substantiated by extensive investigations on the behavior of thin polymer films, as well as empirical observations in traditional composites and filled rubbers.[3] It is expected that the inclusion of nanomaterials will significantly affect the properties of the polymer matrix as well as the polymeric nanocomposites. In ceramic matrix nanocomposites, the motivation of adding nanomaterials is related to the toughness or the crack growth resistance of ceramics considering ceramics are already stiff and strong. Beyond mechanical property concern, there are other objectives of inclusion of nanomaterials in polymer or ceramic matrix, such as increase of thermal conductivity and electronic conductivity. A better understanding of the goal of the nanocomposite and the synthesis-microstructure-property relationships of nanomaterials is of great importance for the success of development of nanocomposites with desired properties.

SiC microfibers are one of the most well studied non-oxide fibers for ceramic matrix composite, especially in harsh environments and at elevated temperatures.[4] SiC nanofibers are expected to have superior thermal stability to CNTs in oxidizing environments. It was reported that SiC nanofibers with 20-30 nm size have tensile strength of over 50GPa, which is about half of the theoretical strength of CNT (100GPa).[5] It appears that SiC nanofibers may be a good candidate as reinforcing materials for various applications. The present study will focus on the synthesis and microstructure characterization of various CNTs and SiC nanofibers fabricated with CVD. It is hoped that the information collected here will help for the design of the composites and selection of reinforcing nanomaterials.

EXPERIMENT
Synthesis of carbon nanotubes
Microwave plasma-enhanced chemical vapor deposition (MPCVD; AsTex 1500i) was used to grow CNTs. The typical growth experiments were conducted with 100 sccm C_2H_2 and 400 sccm NH_3 at about 10 torr. The deposition temperature was varied from 800 to 900 °C and deposition time varied from 5 minutes to 14 hours. Anodic aluminum oxide (AAO) templates (Whatman Inc.) were used for the growth of CNTs. The inner diameter of the pore of the AAO template is about 200 nm. Catalysts were also used for some experiments. In order to prepare catalyst nanoparticles, AAO templates were first immersed in 0.1 M or 1 M $Ni(NO_3)_2$ solution and then put in the CVD chamber and treated with 100 sccm H_2 at 800 °C for about 2 hours. When plasma was employed, the input power was controlled at 1000 W.

Synthesis of SiC nanofibers
A hotwall CVD was used to grow SiC nanofibers. CNTs (Helix Material Solutions, Inc.) with diameter from 30-100 nm were used as the carbon source. A mixture of Si and SiO_2 powder with 2:1 molar ratio was put under the CNTs in a high-purity alumina crucible to generate volatile Si species, which reacted with the CNTs to form SiC nanofibers. The reaction temperature was controlled at approximately 1400 °C. The carrier gas composition was 5% H_2 in Ar.

Characterization

The morphology of the CNTs and SiC nanofibers were examined with SEM (LEO 230) and more detailed information, such as atomic structures, was obtained by TEM (JEOL 2010). All the CNTs fabricated in the present study were embedded in AAO templates. For all the TEM samples and some of the SEM samples, the AAO templates were etched off with KOH solutions at room temperature. No conducting coating was used for any SEM image.

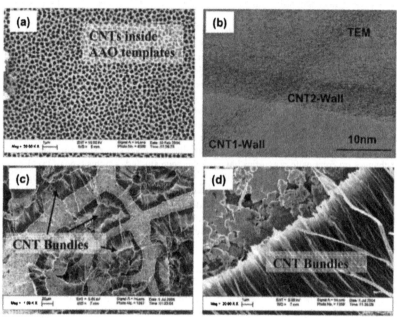

Figure 1. SEM and TEM images of a CNT sample fabricated for 12 hours at 900°C. (a and b) and a CNT sample fabricated for 14 hours at 800 °C (c and d). (a) CNTs in templates, (b) TEM images, (c) and (d) Templates etched off.

RESULTS AND DISCUSSION

Catalyst Free CNTs (No Plasma)

Fabrication of catalyst-free CNTs with such CVD-template method at lower temperatures were studied by Martin's group and Xu's group, but their primary research interests and efforts are not on their mechanical properties.[6,7] The present study will lay emphasis on the correlation of microstructures and mechanical properties. Figure 1 depicts the SEM and TEM images of two samples grown at relatively high temperatures for 12 hours and 14 hours. Both samples turned black after deposition, but not much microstructural difference could be observed before and after deposition based on SEM images (Figure 1a). The deposition rate is very low, only ~0.5nm per hour at 900 °C. CNT graphitic layers are not well ordered, but roughly parallel to the tube axis (Figure 1b). Figure 1c and 1d shows the SEM images of CNT samples when AAO templates were etched off. Most of CNTs formed CNT bundles after matrix was removed. It is also observed that tubular structure of some CNTs collapsed. The SEM and TEM characterization results demonstrate that the CNTs fabricated with the CVD-template methods do not have perfect tubular structures. In a recent study of AAO matrix nanocomposite reinforced by CNTs fabricated by a similar method at 600 °C, the CNT strength is estimated to be about 15-25 GPa, significantly lower than the theoretical strength (~100GPa). The toughness of the AAO templates increase from 0.4 to 5 MPa m$^{-1/2}$ with this type of CNTs based on nanoindentation tests.[8, 9] Although the nanocomposite mechanical properties are not considered excellent, the significant toughness increase is very encouraging. In addition, toughening mechanisms, such as tube pull-out and crack bridging, were observed in such nanocomposites.[8]

The lower tensile strength estimated from the nanoindentation test seems to be attributed to disordered structure of CNT (Figure 1b). A perfect multiwalled CNT may have higher tensile strength, but the very weak van der Waals forces between the tubes will significantly reduce the load transfer. Such phenomenon has been observed experimentally by Cumings and Zettl.[10] In this case, only the outermost layer of multiwalled CNTs may strengthen the matrix. Recent work by Waters et al suggested that that disorder of this type can create more interactions between the graphitic layers in the multiwalled CNTs.[11] While imperfect structures like this may reduce the axial modulus of the CNTs, it should also increase the interlayer shear strength. The effects of these changes on composite properties have not yet been studied, but this type of bonding may be beneficial for load transfer and could thus improve nanocomposite mechanical properties. This suggests a possible tradeoff between multiwalled CNT strength and inter-tube load transfer, such that there may be an optimal level of disorder for multiwalled CNT reinforced nanocomposites. Heat treatments were used to improve the degree the order of graphitic layer, so tailoring such order is possible experimentally. However, further investigation of these effects is required for a better understanding.

Catalyst Assisted Growth of CNTs (No Plasma)
Ni nanoparticles, reduced from Ni(NO$_3$)$_2$, were uniformly deposited on the inner wall of the template pores. In the sample shown in Figure 2, 1 M Ni(NO$_3$)$_2$ was used for catalyst preparation. CNT wall thickness of this sample is in the range of 30-50 nm. It appears that the Ni catalysts significantly increased the deposition rates. The wall thickness is a little smaller when 0.1M Ni(NO$_3$)$_2$ was used for catalyst preparation (images not shown here), but the thickness is still significantly larger than that of catalyst-free samples. More interestingly, secondary CNTs inside the bigger CNT were observed as shown in Figure 2c and illustrated in Figure 2d. The secondary CNTs are curled and tangled inside primary CNTs. Martin et al reported growth of CNT with metal catalyst nanoparticles at lower temperatures around 500-550 °C. At the lower temperature range, the growth of secondary CNTs was impeded and a hollow CNT could be fabricated without forming the secondary CNTs.[6] The secondary CNTs was apparently promoted by a catalyst assisted vapor-liquid-solid (VLS) mechanism. A typical VLS process starts with the dissolution of gaseous reactants into nanosized liquid droplets on a catalyst metal, followed by nucleation and growth of nanotube or nanowire. The one-dimensional growth is induced and dictated by the liquid droplets, whose size remains unchanged in most cases.[12, 13] We speculate that no liquid droplet formed at low temperatures, such as under 550 °C, so that no secondary CNTs could grow via the VLS mechanism. In this case, temperature acts as a switch that can turn on and off the secondary growth of CNTs. Figure 2b shows fracture surface of the same sample. No debonding between the primary CNTs and the

AAO matrix is observed, suggestive of strong interfacial bonding. However the bonding between the primary CNTs and secondary CNTs is relatively weak. Future research will focus on detailed CNT microstructure characterization by TEM and nanocomposites mechanical testing by nanoindentation.

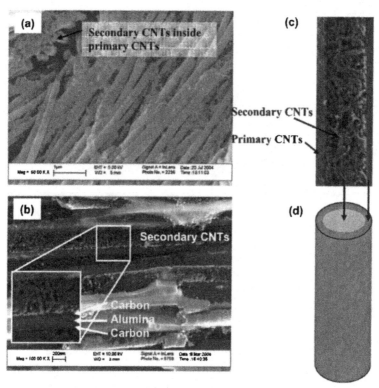

Figure 2. SEM images (a, b, and c) and schematic (d) of a CNT sample grown with Ni nanoparticles catalyst for 1 hour at 800 °C. (a) templates etched off and (b) templates fractured.

Catalyst and Plasma Assisted Growth of CNTs

When both catalysts and plasma was used, extremely high deposition rates were achieved. It is believed the high energy electrons in the plasma dissociate gases and generate atomic and ionized species and thus significantly increase the growth rates. CNT wall thickness reached about 30-50 nm in 5 minutes in the sample shown in Figure 3. CNTs (Figure 3a) did not fall or collapse after templates were etched off. It appears that the CNTs fabricated with plasma are stiffer than those fabricated without plasma. Secondary CNTs were also found in these types of CNTs, but they are less curled compared to those shown in Figure 2 and many of them popped out after AAO matrix was fractured. Both the primary and secondary CNTs are expected to have better structures as indicated by the above observations. It is believed that the high energy electrons contributed to better structures, such as high degree of ordering, of these CNTs. It will be of great value to further investigate the influence of plasma on the structures of CNTs at atomic and meso scales. The secondary CNT popout indicates they are under residual stress inside the primary CNTs and push primary CNTs against the AAO matrix. It is conceivable that the interfacial strength between the primary CNTs and matrix may be improved by the secondary CNTs. Further research will be required to investigate such effects.

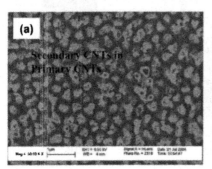

Figure 3. SEM images of CNT grown with catalyst and plasma for 5 minutes at 700 °C. (a) templates etched off and (b) templates fractured.

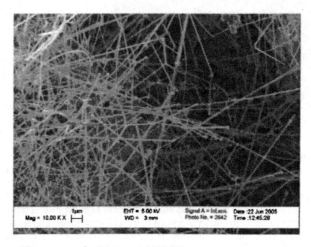

Figure 4. SiC nanofibers grown for 7 hours at 1400 °C.

SiC nanofibers Fabricated in a Hotwall CVD

Figure 4 shows a typical SEM image of SiC nanofibers. The SiC nanofibers grown in a hotwall CVD are straight and appear to be much stiffer than all the CNTs in the present study.

Many small beads could be easily identified at the end of SiC nanofibers, suggestive of vapor-liquid-solid (VLS) growth mechanism. The metal catalysts originated from the original CNT precursor, which may contain 2% of metal nanoparticles or less. The diameters of these SiC nanofibers, ranging from 30-100 nm, match very well with those of CNTs. This is reasonable considering that the sizes of both are controlled by the same nanoparticles. Direct conversion of CNT to SiC is not observed based on SEM and TEM results. The lower end strength of the SiC nanofibers is about 24GPa (based on diameter of 100 nm) and nanofibers with smaller sizes are expected to have higher strength.[14] The average strength of those SiC nanofibers is certainly higher than the CVD catalyst-free multiwalled CNTs fabricated with the plasma off.

The TEM characterization (not shown here) also reveals a 1-5 nm think amorphous coating on the surface of SiC nanofibers. Considering SiC microfibers were usually coated with a coating before added in matrix, this amorphous coating may also serve as an interphase to tailor the bonding strength between the nanofiber and the matrix of interest.

CONCLUSIONS

It is demonstrated that nanomaterials with different microstructures and different nanomaterials could be fabricated for the composite applications with CVD methods. Both the microstructure and the mechanical properties of CNTs could be tailored by modifying the experimental conditions. The deposition rates of catalyst-free CNTs without plasma are extremely low and it may not be practical to use such of CNTs as reinforcing materials despite the improvement of composite toughness. Catalysts and plasma can significantly speed up the deposition rates and plasma may also improve the stiffness of CNTs. Both catalyst and plasma may be desired for fabrication of CNT reinforce AAO nanocomposites. SiC nanofibers with diameters of 30-100 nm were also fabricated with a hotwall CVD. These nanofibers may serve as a good candidate for composite reinforcing materials considering their excellent mechanical properties and superior thermal and chemical resistance.

ACKNOWLEDGEMENTS

This work was supported by the National Science Foundation through a Nanoscale Interdisciplinary Research Team (NIRT) Grant, CMS-0304246.

REFERENCES

[1]W. A. Curtin and B. W. Sheldon, "CNT-reinforced Ceramics and Metals," *Materials Today*, 7, 44-49 (2004).

[2]H. D. Wagner and R. A. Vaia, "Nanocomposites: Issues at the Interface," *Materials Today*, 7, 38-42 (2004).

[3]F. Gao, "Clay/Polymer Composites: The Story," *Materials Today*, 7, 50-55 (2004).

[4]R. Naslain, "Design, Preparation and Properties of Non-Oxide CMCs for Application in Engines and Nuclear Reactors: An Overview," *Compos. Sci. Technol.*, 64, 155-170, (2004).

[5]E. W. Wong, P. E. Sheehan, and C. M. Lieber, "Nanobeam Mechanics: Elasticity, Strength, and Toughness of Nanorods and Nanotubes," *Science*, 227, 1971-75, (1997).

[6]G. Che, B. B. Lakshmi, C. R. Martin, E. R. Fisher, and R. A. Ruoff, "Chemical Vapor Deposition Based Synthesis of Carbon Nanotubes and Nanofibers Using a Template Method," *Chem. Mater.*, 10, 260-67, (1998).

[7]J Li, C. Papadopoulos, J. Xu, M. Moskovits, "Large-Area Hexagonally Ordered Arrays of Carbon Nanotubes," *Appl. Phys. Lett.*, **75**, 367–69, (1999).

[8]Z. Xia, L. Riester, W. A. Curtin, H. Li, B. W. Sheldon, J. Liang, B. Chang, and J. Xu, "Direct Observation of Toughening Mechanisms in Carbon Nanotube Ceramic Matrix Composites," *Acta Mater.*, **52**, 931-944, (2004).

[9]Z. Xia, W. A. Curtin, B. W. Sheldon, "Fracture toughness of highly ordered carbon nanotube/alumina nanocomposites, "*Transactions of the ASME*, **126**, 238–244, (2004).

[10]J. Cumings, and A. Zettl., "Low-Friction Nanoscale Linear Bearing Realized From Multiwalled Carbon Nanotubes," *Science*, **289**, 602-604, (2000).

[11]J. F. Waters, P. R. Guduru,a_ M. Jouzi, J. M. Xu, T. Hanlon and S. Suresh, "Shell Buckling of Individual Multiwalleded Carbon Nanotubes Using Nanoindentation," *Appl. Phys. Lett.*, **87**, 103109, (2005).

[12]R. S. Wagner, W.C. Ellis, "Vapor-liquid-solid mechanism of single crystal growth," *Appl. Phys. Lett.* **4**, 89–90 (1964).

[13]M. Law, J. Goldberger, and P. Yang "Semiconductor nanowires and nanotubes," *Annu. Rev. Mater. Sci.* **34**, 83-122 (2004).

[14]W. Yang, H. Araki, C. Tang, S. Thaveethavorn, A. Kohyama, H. Suzuki, and T. Noda, "Single-Crystal SiC Nanowires with a Thin Carbon Coating for Stronger and Tougher Ceramic Composites," *Adv. Mater.*, **17**, 1519-23, (2005).

3-D MICROPARTICLES OF BaTiO₃ AND Zn₂SiO₄ VIA THE CHEMICAL (SOL-GEL, ACETATE, OR HYDROTHERMAL) CONVERSION OF BIOLOGICAL (DIATOM) TEMPLATES

3-D MICROPARTICLES OF $BaTiO_3$ AND Zn_2SiO_4 VIA THE CHEMICAL (SOL-GEL, ACETATE, OR HYDROTHERMAL) CONVERSION OF BIOLOGICAL (DIATOM) TEMPLATES

Ye Cai, Michael R. Weatherspoon, Eric Ernst, Michael S. Haluska, Robert L. Snyder, and Kenneth H. Sandhage
School of Materials Science and Engineering, Georgia Institute of Technology, Atlanta, GA

ABSTRACT

In this paper, the silica-based microshells of diatoms (planktonic micro-algae) have been used as biologically-replicable (i.e., scalable) 3-D templates for synthesizing functional multicomponent microparticles of controlled shape. Three shape-preserving chemical conversion approaches have been explored for the conversion of diatom microshells (frustules) into functional multicomponent oxides: sol-gel, acetate precursor, and hydrothermal routes. For the sol-gel approach, SiO_2 frustules were first converted by gas/solid reaction into MgO microparticles of similar shape. A conformal and continuous coating of $BaTiO_3$ was then applied, via sol-gel processing, to the MgO microparticles. The underlying MgO templates were then selectively dissolved away to yield $BaTiO_3$ microparticles that retained the overall shape of the starting frustules. In the acetate precursor route, a zinc acetate precursor solution was used to apply a coating of ZnO nanoparticles to SiO_2 frustules. The ZnO nanoparticles were then allowed to react at elevated temperature with the underlying SiO_2 to form Zn_2SiO_4-bearing microparticles with the frustule shape. Finally, a metathetic gas/solid reaction was used to convert SiO_2 frustules into TiO_2. The titania microparticles were then converted into $BaTiO_3$-based microparticles that retained the frustule shape via hydrothermal reaction with a barium hydroxide-bearing solution. These results demonstrate that several shape-preserving chemical approaches may be used to convert biologically-replicable 3-D nanostructured microtemplates into microparticles with functional multicomponent chemistries and with controlled morphologies.

INTRODUCTION

Nature provides spectacular examples of micro-organisms that assemble inorganic nanoparticles into intricate three-dimensional (3-D) microscale structures.[1,2] Although near the bottom of the biological food chain, diatoms (single-celled planktonic algae) assemble silica structures with hierarchical nano-to-microscale features with a mastery that mankind cannot yet match. Each of the tens of thousands of extant diatom species assembles silica nanoparticles into a microshell (frustule) with a distinct 3-D shape and with unique patterns of fine features.[3] The repeated doubling associated with biological reproduction enables these organisms to generate enormous numbers of 3-D nanostructured frustules with similar morphologies (e.g., 40 reproduction cycles can yield more than 1 trillion 3-D replicas).[3,4] The massive parallelism and species-specific (genetic) shape control associated with diatom frustule assembly are very attractive for device manufacturing. However, the silica-based chemistry provides a significant restriction on the range of device applications for diatom frustules. Many, if not most, device applications involving functional ceramics require the use of multicomponent oxide compounds. The objective of this study is to explore several synthetic chemical methods capable of converting silica-based diatom frustules (or other biogenic templates) into multicomponent oxides while preserving the starting frustule shapes.[5,6]

EXPERIMENTAL PROCEDURE

Aulacoseira diatom frustules were selected as biogenic templates in this research. Secondary electron images of the starting *Aulacoseira* frustules are shown in Figure 1. The *Aulacoseira* frustules possessed a hollow cylindrical shape. One end face of each half frustule (Figure 1(a)) contained a circular hole and a protruding outer rim, whereas the other end was closed and exhibited fingerlike extensions. The fingerlike extensions from one half-frustule were observed to interlock with those of another half-frustule to form a paired (complete frustule) assembly, as shown in Figure 1(b). The walls of the frustules also contained rows of fine pores (a few hundred nm in diameter) aligned along the frustule length.

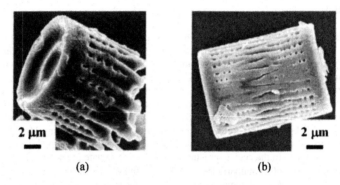

(a)   (b)

Figure 1. Secondary electron images of the as-received *Aulacoseira* diatom frustules. (a) a half-frustule, (b) a complete frustule (i.e., two half-frustules joined end-to-end).

Conformal Sol-Gel (React and Coat) Route to 3-D Frustule-Shaped BaTiO$_3$ Microparticles

Silica-based *Aulacoseira* diatom frustules were first converted into MgO before applying a sol-gel-derived BaTiO$_3$ coating. The shape-preserving magnesiothermic (gas/solid) reaction process used for conversion of the silica frustules into magnesia/silicon composite microparticles has been described elsewhere.[5,7,8] The MgO/Si microparticles were immersed in a 0.49 M NaOH solution with sonication at 60°C for 3 h to selectively dissolve the Si reaction product. The resulting frustule-shaped MgO microparticles were then used as inert templates for the application of a sol-gel-derived BaTiO$_3$ coating. The precursor solution consisted of barium titanium ethylhexanoisopropoxide (Alfa Aesar Chemical Co., Ward Hill, MA) dissolved in absolute ethanol, with smaller additions of deionized water and concentrated (29%) ammonium hydroxide. The BaTi(OOC$_8$H$_{15}$)(OC$_3$H$_7$)$_5$: H$_2$O : NH$_4$OH : EtOH molar ratio in the precursor solution was 1.0 : 0.77 : 0.15 : 110. A 0.4 g batch of the frustule-shaped magnesia microparticles was placed into 20 ml of this solution, and the mixture was stirred and refluxed for 3 h at 70°C. Some of the solution (5 ml) was then allowed to evaporate at 56°C. Refluxing was then repeated for 1 more hour at 70°C. Evaporation of 5 more ml of the solution was then conducted at 56°C. After a final 70°C refluxing treatment for an hour, the solution was allowed to completely evaporate at 56°C. The coated frustule-shaped particles were heated in air at 3°C/min in air to

700°C, and then held at this temperature for 1.5 h to allow for conversion of the coating into BaTiO₃. The underlying MgO scaffold was then removed from the conformal BaTiO₃ coating by immersion of the frustule-shaped particles in a magnetically-stirred 0.7 M HCl solution at 60°C for 1 h.

Conformal Acetate (Coating and React) Route to 3-D Frustule-Shaped Zn₂SiO₄-bearing Microparticles

In this case, the reaction between a ZnO coating and the underlying SiO₂ frustules was used to generate Zn₂SiO₄-bearing frustules. *Aulacoseira* diatom frustules (0.2 g) were added to a 30 mL solution comprised of 40 vol% ethanol in water. After heating the mixture to 90°C and stirring (300 rpm) for 10 min, triethanolamine (reagent grade, Alfa Aesar) was introduced at a concentration of 1.6 mol/L. Zinc acetate dihydrate (reagent grade, Alfa Aesar) was then added to the mixture to achieve a concentration of 0.04 mol/L. The mixture was then stirred for 2 more h at 90°C. The frustules were then removed from the solution by filtration and fired at 600°C for 4 h in air. The coated frustules were heat treated for 4 h at 1100°C in air to allow for the formation of Zn₂SiO₄ by the reaction of the ZnO coating with the underlying SiO₂.

Hydrothermal (Serial Reaction) Route to 3-D Frustule-Shaped BaTiO₃-based Microparticles

The *Aulacoseira* diatoms were first converted into titania by a shape-preserving metathetic (gas/solid) reaction process described elsewhere.[5,7,9] A 50 mg batch of the frustule-shaped titania microparticles was loaded along with 1.75 g of barium hydroxide octahydrate (Ba(OH)₂·8H₂O, >98% purity, Alfa Aesar) and 1.75 mL of deionized water into a 3 mL PTFE tube within an Ar atmosphere glove box. The tube was then sealed and removed from the controlled atmosphere environment and placed in an oven at 90°C for 24 h. The resulting reacted frustules were then removed from the tube, thoroughly washed with deionized water at 80°C, and then dried at ~ 80°C for 12 h.

RESULTS AND DISCUSSION
Conformal Sol-Gel (React and Coat) Route to 3-D Frustule-Shaped BaTiO₃ Microparticles

BaTiO₃ is not chemically compatible with SiO₂; that is, a number of stable ternary compounds (e.g., BaTiSiO₅, BaTiSi₂O₇, BaTiSi₃O₉) can form by the reaction of these two oxides.[10] Indeed, initial attempts to produce BaTiO₃ on SiO₂ via low-temperature firing of sol-gel-derived coatings yielded BaTiSiO₅. BaTiO₃ and MgO are, however, chemically compatible (i.e., a tie line exists between these oxides).[11] Hence, in order to generate a sol-gel-derived frustule-shaped BaTiO₃ microparticle, it was first necessary to convert the SiO₂ frustules into frustule-shaped MgO microparticles. A secondary electron image of a frustule-shaped MgO microparticle, produced by magnesiothermic reduction of silica and then selective dissolution of the silicon product, is shown in Figure 2(a). The fine MgO nanocrystals that comprise the chemically-converted frustule can be seen in Figure 2(a). This image indicates that the shape and features of *Aulacoseira* frustules (e.g., fingerlike extensions; rows of pores along the frustule wall) were retained in the MgO microparticle. A secondary electron image of a freestanding, frustule-shaped BaTiO₃ microparticle, produced by firing a sol-gel-coated MgO frustule at 700°C for 1.5 h and then selectively dissolving away the MgO scaffold, is shown in Figure 2(b). Energy dispersive x-ray (EDX) analysis of the microparticle shown in Figure 2(b) is presented in Figure 2(c). The presence of the Ba, Ti, and O peaks, and the absence of the Mg peak at 1.25 eV, in the EDX pattern confirmed that the magnesia template was completely and selectively removed from the conformal BaTiO₃ coating by the acid treatment (note: the Al, Au, and C peaks in this EDX pattern were generated by the aluminum SEM stub, the Au coating applied to prevent charging, and the C tape used to provide a conduction path, respectively). X-ray diffraction (XRD) analysis also confirmed the presence and absence of BaTiO₃ and MgO, respectively, after the selective dissolution treatment. Figure 2(b) clearly demonstrates that this combined reaction and conformal sol-gel coating approach can be used to convert SiO₂-based diatom frustules into freestanding BaTiO₃ structures that preserve the starting frustule shape and fine features.

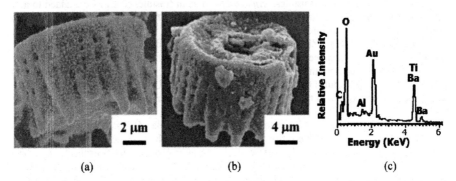

(a)   (b)   (c)

Figure 2. Secondary electron images of *Aulacoseira* diatom frustules after shape-preserving conversion into: a) MgO, and b) freestanding BaTiO₃. (c) EDX analysis of the converted frustule-shaped BaTiO₃ microparticle shown in (b).

Conformal Acetate (Coating and React) Route to 3-D Frustule-Shaped Zn₂SiO₄-bearing Microparticles

A secondary electron image of an *Aulacoseira* frustule after exposure to the zinc acetate-bearing precursor solution and then firing at 650°C for 4 h is shown in Figure 3(a). The frustules were covered with a porous coating of nanoparticles after this treatment. XRD analysis (Figure 3(d)-1) indicated that these nanoparticles were comprised of ZnO. Measurement of the widths of the ZnO diffraction peaks at half maximum and use of the Scherrer equation yielded an average ZnO crystallite size of only 20 nm. The coated frustules were then heated for 4 h at 1100°C in ambient air in order to allow the ZnO nanoparticles to undergo the following solid-state reaction with the underlying SiO₂ frustules (note: Zn₂SiO₄ is the only stable solid compound at ambient pressures that can form by the reaction of ZnO with SiO₂[12]):

$$2ZnO + SiO_2 \rightarrow Zn_2SiO_4 \tag{1}$$

XRD analysis (Fig. 3(d)-2) indicated that this 1100°C/4 h treatment resulted in the consumption of ZnO to yield Zn₂SiO₄ (willemite). Figure 3(b) reveals a secondary electron image of the resulting reacted frustules, which retained the general *Aulacoseira* frustule morphology. A low magnification transmission electron (TE) image of a cross-section of such a reacted frustule is shown in Figure 3(c). After the 1100°C/4 h treatment, the porous coating of fine ZnO particles was converted into a dense, continuous, and conformal Zn_2SiO_4 coating on the external surfaces of the silica frustules.

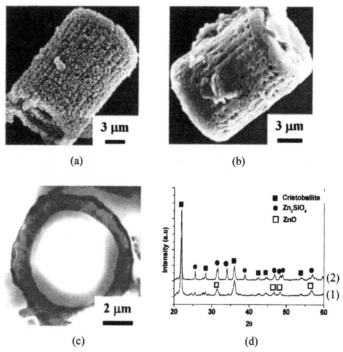

Figure 3. Secondary electron images of *Aulacoseira* diatom frustules after: (a) immersion in a zinc acetate solution with ethanol, water, and triethanolamine at 90°C for 2 h, and then heat treatment at 650°C for 4 h, and (b) after further heat treatment at 1100°C for 4 h. (c) TEM micrograph of a frustule cross-section showing the continuous zinc silicate (relatively dark) coating on the frustule surface. (d) X-ray diffraction analyses of the *Aulacoseira* frustules: (1) after immersion in a zinc acetate solution with ethanol, water, and triethanolamine at 90°C for 2 h, and then heat treatment at 650°C for 4 h, and (2) after further heat treatment at 1100°C for 4 h.

Hydrothermal (Serial Reaction) Route to 3-D Frustule-Shaped BaTiO₃-based Microparticles

A secondary electron image of a frustule-shaped titania microparticle generated by a metathetic (gas/solid) reaction process is shown in Figure 4(a).[5,7,9] It can be seen that the shape and features of the starting *Aulacoseira* diatom frustule were preserved in the titania microparticle. Preliminary experiments indicated that such titania frustules could undergo hydrothermal reaction at 150°C with barium hydroxide-bearing solutions.[13] Further work has indicated that such reactions may be conducted below 100°C. A reacted frustule generated after hydrothermal exposure of such titania microparticles to a barium hydroxide solution for 24 h at 90°C is shown in Figure 4(b). The *Aulacoseira* frustule morphology was preserved in this reacted frustule. XRD analysis (Figure 4(c)) confirmed that this hydrothermal treatment resulted in conversion of the titania microparticles into BaTiO₃, with very little residual unreacted TiO₂ or BaCO₃ (the latter phase is likely to have formed by the reaction of a small amount of barium hydroxide or barium hydroxide octahydrate with carbon dioxide upon exposure to ambient air).

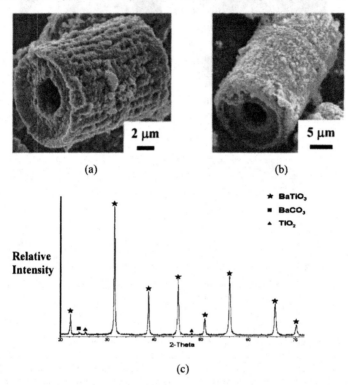

(a)   (b)

(c)

Figure 4. Secondary electron images of *Aulacoseira* diatom frustules after conversion into: a) a frustule-shaped titania microparticle via a metathetic (gas/solid) process, and b) a BaTiO₃-based microparticle by subsequent hydrothermal reaction with a barium hydroxide-bearing solution. c) XRD analysis of the frustule-shaped BaTiO₃-based microparticle after hydrothermal reaction.

CONCLUSIONS

Three synthetic chemical approaches have been developed for converting the silica-based composition of diatom frustules into multicomponent oxides (BaTiO₃, Zn₂SiO₄) while preserving the starting 3-D frustule morphology. Freestanding frustule-shaped BaTiO₃ microparticles were generated by applying a sol-gel-derived conformal BaTiO₃ coating to MgO-converted frustules, and then removing the underlying MgO by selective acid dissolution.

Zn$_2$SiO$_4$-bearing frustules were produced by using an acetate precursor solution to apply a ZnO nanoparticle coating to the frustules, and then allowing the nanoparticles to react with the underlying SiO$_2$. Frustule-shaped BaTiO$_3$-based microparticles were also generated through the hydrothermal reaction of titania microparticles (formed by a metathetic displacement reaction with silica frustules) with a barium hydroxide-bearing solution.

The successful demonstration of such shape-preserving react-and-coat, coat-and-react, and serial reaction conversion processes indicates that diatom frustules (or other biomineralized structures) may be converted into a wide variety of other functional multicomponent chemistries with a retention of the biologically-derived shapes. This synergistic combination of biological assembly with synthetic chemical functionalization opens the door to large numbers of low cost nanostructured microparticles with well-controlled 3-D shapes, chemistries, and properties that can be tailored for a host of applications.[5,6]

ACKNOWLEDGEMENTS
The financial support provided by the Air Force Office of Scientific Research (Dr. Joan Fuller and Dr. Hugh C. De Long, Program Managers) is gratefully acknowledged.

REFERENCES
[1]H. A. Lowenstam, "Minerals Formed by Organisms," *Science*, **211**, 1126–1131 (1981).

[2]E. Bauerlein, "Bomineralization of Unicellular Organisms: An Unusual Membrane Biochemistry for the Production of Inorganic Nano- and Micro- structures," *Angew. Chem. Int. Ed.*, **42**, 614–641 (2003).

[3]F. E. Round, R. M. Crawford, and D. G. Mann, *The Diatoms: Biology and Morphology of the Genera*. Cambridge University Press, Cambridge, England, 1990.

[4]S. A. Crawford, M. J. Higgins, P. Mulvaney, and R. Wetherbee, "Nanostructure of the Diatom Frustule as Revealed by Atomic Force and Scanning Electron Microscopy," *J. Phycol.*, **37**, 543–554 (2001).

[5]K. H. Sandhage, "Shaped Microcomponents via Reactive Conversion of Biologically-derived Microtemplates," *U.S. Patent Application*, pending.

[6]C. S. Gaddis, K. H. Sandhage, "Process for Fabricating Microcomponents," *U.S. Patent Application*, pending.

[7]K. H. Sandhage, M. B. Dickerson, P. M. Huseman, M. A. Caranna, J. D. Clifton, T. A. Bull, T. J. Heibel, W. R. Overton, and M. E. A. Schoenwaelder, "Novel, Bioclastic Route to Self-Assembled, 3-D, Chemically Tailored Meso/Nanostructures: Shape-Preserving Reactive Conversion of Biosilica (Diatom) Microshells," *Adv. Mater.*, **14**, 429–433 (2002).

[8]Y. Cai, S. M. Allan, and K. H. Sandhage, "Three-dimensional Magnesia-based Nanocrystal Assemblies via Low-temperature Magnesiothermic Reaction of Diatom Microshells," *J. Am. Ceram. Soc.*, **88**, 2005-1010 (2005).

[9]R. R. Unocic, F. M. Zalar, P. M. Sarosi, Y. Cai, and K. H. Sandhage, "Anatase Assemblies from Algae: Coupling Biological Self-Assembly of 3-D Nanoparticle Structures with Synthetic Reaction Chemistry," *Chem. Comm.*, **7**, 795–796 (2004).

[10]D. E. Rase and R. Roy, "Phase Equilibria in the System Barium Titanate–Silica," *J. Am. Ceram. Soc.*, **38**, 389–395 (1955).

[11]R. S. Roth, W. S. Brower, M. Austin, and M. Koob, "System BaO–MgO–TiO$_2$," *Phase Diagrams for Ceramists*, Vol. VI, eds. R. S. Roth, J. R. Dennis, and H. F. McMurdie. The American Ceramic Society, Westerville, OH, 264,1987.

[12]E. N. Bunting, "Phase Equilibria in the System: SiO$_2$-ZnO," *J. Am. Ceram. Soc.*, **13**, 5-10 (1930).
[13]C. S. Gaddis, "Diatom Alchemy," *M.S. Thesis*, Georgia Institute of Technology, 2004.

POLYMER FIBER ASSISTED PROCESSING OF CERAMIC OXIDE NANO AND SUBMICRON FIBERS

Satyajit Shukla, Erik Brinley, Hyoung J. Cho, and Sudipta Seal

Surface Engineering and Nanofabrication Facility (SNF)
Advanced Materials Processing and Analysis Center (AMPAC) and
   Mechanical Materials Aerospace Engineering (MMAE) Department
Engineering 381
4000 Central Florida Blvd.
Orlando, Florida 32816
Phone: 407-882-1189
Fax: 407-882-1462
E-mail(s): sseal@mail.ucf.edu, sshukla@ucf.edu

ABSTRACT

Nano and Submicron ceramic oxide fibers of tin oxide (SnO_2) have been processed using the polymer fibers as templates. Highly porous fibrous mat of hydroxypropyl cellulose (HPC) polymer (molecular weight 80000 g/mol) has been obtained via electrospinning technique. In order to derive the ceramic oxide fibers using the polymer fibers as templates, the electrospinning characteristics of the HPC polymer has been established by varying the critical processing parameters such as polymer concentration, solvent type, tip-to-electrode distance, feeding speed, and applied voltage. Under selected processing conditions, the Sn-precursor is mixed with the polymer solution in alcohol, and then, the electrospun porous HPC polymer mat is converted to porous ceramic fiber network by using a suitable calcination treatment. Nano and submicron fibers of SnO_2 have been deposited on the microelectromechanical system (MEMS) device, consisting oxidized silicon (Si/SiO_2) wafer with gold (Au) electrode pattern. Such micro-device is highly suitable for the room temperature gas (typically hydrogen) sensing application. In conclusion, very effective polymer assisted processing of ceramic oxide nano and submicron fibers have been demonstrated for the low temperature gas sensing application.

INTRODUCTION

Electrospinning is a technique to produce the nano and the submicron sized polymer fibers.[1] A highly dense polymer solution is utilized for this purpose, which is placed in a syringe under the application of an electric field. If sufficiently high voltage is applied in between the syringe and the collector, then a polymer jet is ejected from the tip of the syringe. The polymer fibers are then deposited on the grounded substrate. Various electrospinning parameters, such as the molecular weight and the concentration of the polymer, the solvent type, the solution viscosity and the conductivity, the surface tension, the tip-to-collector distance, the applied voltage, the flow rate, and the surrounding atmosphere, significantly affect the characteristics of the electrospun polymer fibers.[2]

Although fibers of different polymers are formed via electrospinning technique, the electrospinning characteristics of the hydroxypropyl cellulose (HPC) polymer have not been investigated in the literature, which has been utilized earlier in the powder form, as a steric stabilizer, in the synthesis of nanocrystalline ceramic oxide powders.[3] In this investigation, we demonstrate the use of electrospinning technique in forming the HPC polymer fibers and utilize them as templates in the synthesis of semiconductor tin oxide (SnO_2) fibers.

EXPERIMENTAL

Anhydrous ethanol was purchased from Alfa Aesar (U.S.A.) and anhydrous 2-propanol, HPC polymer (molecular weight 80000 g/mol) and tin(II) chloride ($SnCl_2$) were supplied by Sigma-Aldrich (U.S.A.). All the chemicals were used as received.

The experimental set-up consists of a syringe, syringe pump, Cu-plate, Al-foil, and a high voltage power supply (0-30 kV). The syringe having a metallic tip holds the viscous polymer solution and is mounted over the syringe pump, which maintains the flow of the polymer solution to the tip of the syringe at a constant rate. The Cu-plate covered with Al-foil is placed in front of the syringe tip to collect the electrospun HPC nano and submicron fibers by adjusting the tip-to-collector distance as desired. The positive terminal of the high voltage power supply is connected to the metallic syringe-tip while the negative (ground) terminal is connected to the Cu-plate covered with Al foil.

Proper amount of the HPC polymer (15 wt%) was weighed and dissolved completely in the appropriate solvent under continuous stirring using the magnetic stirrer. The beaker containing the polymeric solution was covered with a paraffin tape during stirring. Proper amount of HPC solution is taken in the syringe, which is then placed on the syringe pump. After applying the desired electric potential in between the metallic syringe-tip and the Cu-plate, the HPC solution is continuously fed to the syringe-tip at the constant flow rate of 20 μl/min using the syringe pump. During the electrospinning process, the Al-foil gradually changed its color from grey to white indicating the deposition of the HPC fibers. The electrospinning of the HPC polymer dissolved in anhydrous ethanol and 2-propanol was conducted for the two different tip-to-collector distances, 10 cm and 15 cm, by using 15 kV, 25 kV, and 30 kV applied voltages.

To synthesize SnO_2 fibers using the electrospun HPC fibers, 0.45 M of $SnCl_2$ was dissolved completely in 15 wt% HPC solution in ethanol. The tip-to-collector distance of 10 cm was selected with the applied potential difference of 15 kV. An oxidized silicon (Si/SiO_2) wafer having the patterned gold (Au) electrodes was placed on the Al foil to collect the HPC fibers with the Sn-precursor. After the electrospinning process, the Si/SiO_2 wafer was subjected to the thermal treatment at 700 °C using the heating rate of 30 °C/min and holding time of 1 h. The substrate was then cooled down to room temperature inside the furnace.

The morphology and the average diameter of the electrospun HPC and SnO_2 fibers were analyzed using the scanning electron microscopy (SEM) (JSM-6400F, JEOL, Tokyo, Japan). To avoid any surface charging during the SEM analysis, the fibers were sputter coated with 30 nm Au–Pd layer. The energy dispersive analysis of x-rays (EDX) was performed to study the bulk chemistry of the fibers. X-ray diffraction (XRD, Rigaku, Japan) analysis was performed using the Cu Kα radiation (wavelength=1.5418 A).

RESULTS

The SEM images of the HPC fibers obtained using the ethanol solution, with the tip-to-collector distance of 10 cm and the applied voltages of 15 kV, 25 kV, and 30 kV, are presented in Figs. 1(a)-(c) respectively at low magnification. The HPC fibers are formed with beads under these processing conditions with typical bead-on-string morphology. The average bead size and

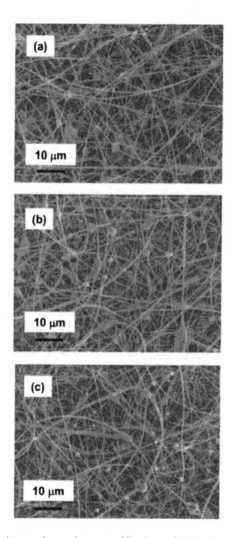

Fig. 1. SEM micrographs, at low magnification, of HPC fibers obtained via electrospinning using anhydrous ethanol as a solvent and the tip-to-collector distance of 10 cm. In (a), (b), and (c), the applied voltage is 15 kV, 25 kV, and 30 kV respectively.

their number density are observed to be a function of the applied voltage. With increasing applied voltage, the average bead size is noted to increase with the bead number density, Figs. 1(a) and 1(b). With further increase in the applied voltage, Fig. 1(c), the bead number density increases; however, the average bead size is qualitatively noted to decrease. The SEM images of the HPC fibers obtained under high magnification are presented in Fig. 2. It is clearly noted that, the average HPC fiber diameter increases first with increasing applied voltage, Figs. 2(a) and 2(b), and then decreases, Fig. 2(c).

The SEM image of the HPC fibers obtained using the anhydrous ethanol solution, with the tip-to-collector distance of 15 cm and the applied voltages of 15 kV, 25 kV, and 30 kV are presented in Figs. 3(a)-(c) respectively. It is qualitatively observed that, the average HPC fiber diameter, the average bead size, and their number density increase with increasing applied potential difference. However, the comparison of Figs. 1 and 3 shows that, for the tip-to-collector distance of 15 cm, no decrease in the average fiber size and average bead size is noted at the highest applied voltage of 30 kV.

The SEM images of the HPC fibers obtained using the 2-propanol solution, with the tip-to-collector distance of 10 cm and the applied voltages of 15 kV, 25 kV, and 30 kV are presented in Figs. 4(a)-(c) respectively at low magnification. In contrast to the bead-formation tendency with the HPC solution in ethanol, Fig. 1, the HPC fibers are formed without any beads under all processing conditions when 2-propanol is used as a solvent. It is also qualitatively noted that, the average HPC fiber diameter increases first with increasing applied voltage, and then, it decreases. This variation is similar to the one that is observed in Fig. 2.

Typical EDX analysis of the HPC fiber is presented in Fig. 5(a), where a strong C-peak is seen, which is originating from the HPC polymer (Note: Au and Pd peaks are due to sputtering process). The EDX spectra obtained from the calcination treatment of the HPC fibers with Sn^{2+} and Cl^- ions is presented in Fig. 5(b). Very strong C-peak, as observed in Fig. 5(a), has been reduced in intensity significantly after the calcinations treatment. The presence of Sn-peak is, however, noted with the absence of Cl-peak. This possibly suggests the elimination of Cl-peak from the bulk of the fibers after the calcinations treatment.

The SEM image of the SnO_2 fibers deposited on the Si-wafer with the Au-electrodes is shown in Fig. 6(a). Highly porous fibrous network of SnO_2 fibers short circuiting the Au electrodes is observed in Fig. 6(a). The XRD spectra of SnO_2 fibers is presented in Fig. 6(b), where (110), (101), and (211) peaks of tetragonal crystal structure are identified by comparison with the JCPDS data file # 41-1445.

DISCUSSION

The HPC fibers are successfully synthesized using the granular HPC powder via electrospinning technique. According to the proposed mechanism[4,5] of the fiber formation via electrospinning, the polymer solution held in the syringe forms a droplet with flat surface, which is held at the syringe tip by the surface tension force. When the high voltage is applied in between the collector and syringe tip, the droplet surface is pulled into the approximate shape of a section of a sphere by the electric forces and the surface tension. As the bulge forms, the charges produced by the applied electric field, move through the polymer solution to concentrate in the area that is protruding the most. The accumulation of the electric charges causes the droplet surface to protrude more, and since the charge per unit surface area is highest near the greatest protrusion, the surface is pulled into a conical shape, which is known as a Taylor cone.

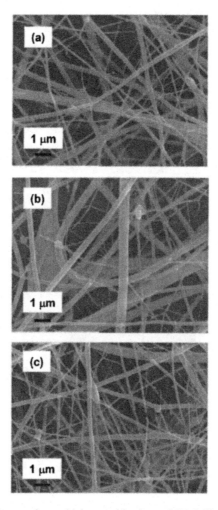

Fig. 2. SEM micrographs, at high magnification, of HPC fibers obtained via electrospinning using anhydrous ethanol as a solvent and the tip-to-collector distance of 10 cm. In (a), (b), and (c), the applied voltages are 15 kV, 25 kV, and 30 kV respectively.

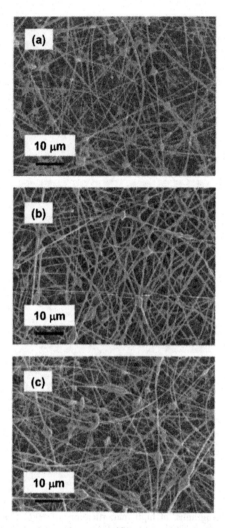

Fig. 3. SEM micrographs, at low magnification, of HPC fibers obtained via electrospinning using anhydrous ethanol as a solvent and the tip-to-collector distance of 15 cm. In (a), (b), and (c), the applied voltages are 15 kV, 25 kV, and 30 kV respectively.

Fig. 4. SEM micrographs, at low magnification, of HPC fibers obtained via electrospinning using anhydrous 2-Propanol as a solvent and the tip-to-collector distance of 10 cm. In (a), (b), and (c), the applied voltages are 15 kV, 25 kV, and 30 kV respectively.

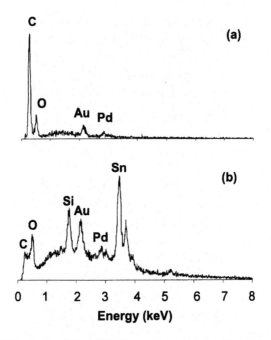

Fig. 5. EDX analysis of the HPC fibers (a), and SnO_2 fibers (b), which are obtained after calcining the HPC fibers, containing the Sn^{2+} and Cl^- ions, at 700 °C for 1 h.

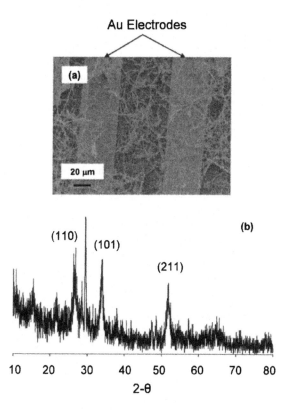

Fig. 6. (a) SEM image of the SnO₂ fibers over the Si/SiO₂ substrate having the patterned Au electrodes, (b) XRD spectrum of the SnO₂ fibers.

The charge per unit area at the tip of the cone increases as the radius near the tip of the cone decreases. As the applied potential is increased, a jet of charged polymer solution is pulled from the syringe tip and the electrospinning process begins. If the applied potential difference is increased further, multiple jets may be ejected from the Taylor cone.[6] As the jet travels from the base to the collector, its diameter decreases and the length increases in a way that keeps constant mass of charged polymer jet per unit time passing any point on the axis. The electric charges, in the form of ions, tend to drift in response to the applied electric field, and they transfer the forces from the electric field to the polymer mass. As the electric charges move to the grounded terminal, they complete the electrical circuit, which provides the energy needed to accelerate the polymer. As the jet moves forward, the solvent gets evaporated depending on the surface tension of the polymer solution. This causes the jet diameter to decrease faster. As the jet diameter decreases with the simultaneous increase in its length, the electric charges are brought closer, and as a result, they start repelling each other. This causes the jet to expand in radial direction and stretch along its axis. Upon closer approach of the electric charges, the radial force become stronger than the cohesive strength of the charged polymer jet, and as a consequence, a single charged polymer jet splits into two or more jets, which is known as 'splaying'.[1,7] This jet division may take place several times in rapid succession producing large number small electrically charged fibers moving towards collector. The charge polymer fibers are, thus, deposited on the grounded substrate. Since the deposited polymer fibers are charged, they repel the polymer fibers, which are deposited subsequently. Hence, the deposited polymer fibers do not overlap and produce a highly porous non-woven polymer fibers mat, as observed in Figs. 1-4.

The HPC fibers with beads are formed with a typical 'bead-on-string' morphology when anhydrous ethanol is used as a solvent instead of 2-propanol, which is similar to the observation reported earlier for other polymers.[2,7,8] During the electrospinning, the beads are basically formed if the forces such as viscoelastic, charge-repulsion, and electrostatic attraction, which tend to elongate the charged polymer jet are overcome by the surface tension force, which tend to break up and spherodize the charged polymer jet. The formation of beads with anhydrous ethanol, hence, suggests that the surface tension force on the charged polymer jet is not completely overcome by the combination of all counter forces with ethanol as a solvent. This also indicates that, the HPC solution in 2-propnaol has higher solution viscosity and lower surface tension relative to those of the HPC solution in ethanol.

Further, when ethanol is used as a solvent, it is noted that, the average HPC fiber diameter and the average bead size increase first with increasing applied voltage, and then decrease. The charge concentration within the polymeric solution increases with increasing applied voltage. As a result, more amount of charged polymer is possibly ejected from the syringe, which increases the average HPC fiber diameter. However, at higher applied voltage, the multiple jet formation and the splaying may take place, which not only reduce the average HPC fiber diameter but also the average bead size. Such effect is, however, only observed when the tip-to-collector distance is 10 cm and not for the working distance of 15 cm. It is likely that, the reduced electric field strength at higher working distance is not able to cause the multiple jet formation and the splaying to occur effectively, resulting in gradual increase in the average HPC fiber diameter and the average bead size with increasing applied potential difference. A detail discussion (qualitative and quantitative) on the effect of applied voltage and tip-to-collector distance on the average polymer (including HPC) fiber diameter and the average bead size could be found elsewhere.[9,10]

In the present investigation, SnO_2 fibers have been produced using the HPC fibers as templates. For this, Sn^{2+} and Cl^- ions have been mixed with the HPC solution in ethanol and HPC/Sn^{2+}/Cl^- composite fibers have been obtained. Using higher calcination treatment under the normal atmospheric conditions, the HPC polymer is decomposed with simultaneous removal of Cl^- ions, which results in the formation crystalline SnO_2 fibers having the tetragonal crystal structure. Moreover, the porous structure of the SnO_2 fibers short circuits the Au electrodes, which are patterned on the Si/SiO_2-wafer; this in turn suggests the semiconducting nature of the SnO_2 fibers. Such a design is highly suitable for the gas sensing application and work is in progress to measure the H_2 sensitivity of the SnO_2 fibers derived using the electrospinning process.

CONCLUSIONS

The HPC polymer fibers have been synthesized via electrospinning technique using two tip-to-collector distances and two different solvents with increasing applied voltage within the range of 10-30 kV. The HPC fibers with beads are obtained when ethanol is used as a solvent; however, the beads are eliminated using the HPC solution in 2-propanol. The average HPC fiber diameter and the average bead size are observed to be function of the applied voltage and the tip-to-collector distance, and they follow the trend which is well described by the known electrospinning mechanism.

ACKNOWLEDGEMENTS

The authors thank UCF, Florida Space Grant Consortium (FSGC), NASA-Glenn (NASA NAG 32751), KSC-NASA, National Science Foundation (NSF EEC–0136710 and NSF CTS 0350572) (Seal) and (NSF CAREER, ECS-0348603 (Cho)) for funding the sensor and the nano-technology research.

REFERENCES

[1]D. Reneker and I. Chun, "Nanometre Diameter Fibers of Polymer Produced by Electrospinning", *Nanotechnology,* 7, 216-23 (1996).

[2]Z. Huang, Y. Zhang, M. Kotaki, and S. Ramakrishna, "A Review on Polymer Nanofibers by Electrospinning and Their Applications in Nanocomposites", *Compos. Sci. Technol.,* 63, 2223-53 (2003).

[3]S. Shukla, S. Seal, and R. Vanfleet, "Sol-Gel Synthesis and Phase Evolution Behavior of Sterically Stabilized Nanocrystalline Zirconia", *J. Sol-Gel Sci. Technol.,* 27, 119-36 (2003).

[4]D. Reneker, A. Yarin, H. Fong, and S. Koombhongse, "Bending Instability of Electrically Charged Liquid Jets of Polymer Solutions in Electrospinning", *J. Appl. Phys.,* 87, 4531-47 (2000).

[5]Y. Shin, M. Hohman, M. Brenner, and G. Rutledge, "Experimental Characterization of Electrospinning: the Electrically Forced Jet and Instabilities", *Polymer,* 42, 09955-67 (2001).

[6]M. Demir, I. Yolgor, E. Yilgor, and B. Erman, "Electrospinning of Polyeurethene Fibers", *Polymer,* 43, 3303-09 (2002).

[7]J. Deitzel, J. Kleinmeyer, D. Harris, and N. Tan, "The Effect of Processing Variables on the Morphology of Electrospun Nanofibers and Textiles", *Polymer,* 42, 261-72 (2001).

[8]H. Fong, I. Chun, and D. Reneker, "Beaded Nanofibers formed During Electrospinning", *Polymer,* 40, 4585-92 (1999).

[9]S. Shukla, E. Brinley, and S. Seal, "Electrospinning of Hydroxypropyl Cellulose Polymer Fibers and Their Application in Synthesis of Nano and Submicron Tin Oxide Fibers", *Polymer* **46**, 12130-145 **(2005)**.

[10]K. Lee, H. Kim, H. Bang, Y. Jung, and S. Lee, "The Change of Bead Morphology Formed on Electrospun Polystyrene Fibers", *Polymer* **44**, 4029-34 **(2003)**.

PHASE DEVELOPMENT IN THE CATALYTIC SYSTEM V$_2$O$_5$/TiO$_2$ UNDER OXIDIZING CONDITIONS

D. Habel, E. Feike, C. Schröder, H. Schubert
Institute for Material Science and -technologies, TU Berlin
Englische Straße 20
D- 10587 Berlin, Germany

A. Hösch
Institute for Applied Geosciences, TU Berlin
Straße des 17. Juni 135
D- 10669 Berlin, Germany

J.B. Stelzer, J. Caro
Institute for Physical Chemistry & Electrochemistry, University of Hanover
Callinstraße 3 - 3a
D- 30167 Hanover, Germany

C. Hess, A. Knop-Gericke
Fritz-Haber-Institute of the Max Planck Society, Department of Inorganic Chemistry,
Faradayweg 4 –6
D-14195 Berlin, Germany

ABSTRACT
    The target of this work was to investigate phase development in the catalyst system consisting of TiO$_2$ (Anatase) and V$_2$O$_5$ (Shcherbinaite). Thus a set of V$_2$O$_5$/TiO$_2$ specimens was prepared by ball milling and exposed to subsequent annealing in air in the temperature range from 400 to 700 °C. The XRD-results showed the presence of Anatase and Shcherbinaite as the only phases up to 525 °C. For temperatures above 525 °C Rutile as a new TiO$_2$-phase occured. Peak intensities and positions were shifted. No loss of oxygen or vanadium was detected. The reaction involves the formation of a Rutile solid solution containing VO$_x$ species. XPS studies showed an oxidation state of 4.75 for V in the Rutile solid solution as compared to 4.65 in the Shcherbinaite. The Rutile solid solution was first found at 525 °C < T < 550 °C for compositions of 3 mole % < V$_2$O$_5$ < 5 mole %. The phase field for Rutile solid solutions extends to 10 mole%< V$_2$O$_5$ <12.5 mole % at 675 °C. For very high V$_2$O$_5$ concentrations an eutectic reaction was found at 631 °C (95 mole% V$_2$O$_5$). A Shcherbinaite structure remained with shifted peak intensities and positions due to the alloying of Ti-ions.
SEM inspections showed that the Rutile formation and the eutectic reaction both cause a substantial grain growth and a loss of surface area. The catalytic activity is entirely lost. The knowledge of phase relations helps to find the appropriate processing conditions and to understand the aging phenomena of catalysts.

INTRODUCTION
    Commonly used catalysts in industry are composed of highly dispersed particles of precious metals or transition metal oxides on oxide supports [1]. Both the particle size and the form of

connection between the support and the active compound have a strong influence on the activity and selectivity of the catalysts [2, 3].

The tailoring of catalysts with defined properties requires deep fundamental knowledge about their operation and aging or deactivation which can only be obtained reliably when the catalysts are monitored under operating conditions. The catalyst may decay in many ways [4-9]. The principle effects of deactivation of catalysts are mainly caused by deposition, sintering, contamination or decomposition. Nevertheless, these can be grouped basically into five intrinsic mechanisms of catalyst decay: (1) poisoning, (2) fouling, (3) thermal degradation, (4) loss of catalytic phases by vapor compound formation accompanied by transport and (5) attrition. (1) and (4) are basically chemical in nature, whereas (2) and (5) are mechanical. There has been reported numerous applications of the $TiO_2 - V_2O_5$ system [10-15]. Phase reaction was investigated by Bond et al. [13] who found Rutile formation under feed gas conditions associated with a loss of oxygen and a predominantly 4 valent V-ion. The change of the metastable Anatase to Rutile is the subject of the work of Pask who found that contaminants reduce the transformations temperature [16].

This work is focussed on the phase development during catalyst processing under oxidizing conditions. The oxidative dehydrogenation of propane (ODP) was selected as a praxis-relevant model reaction for the evaluation of the catalytic properties of the catalysts thus treated. The resulting XRD spectra were qualitatively used to construct a preliminary phase diagram. The measurements were completed by DTA and SEM investigations.

EXPERIMENTAL

Sample preparation for phase analysis

A pure commercial TiO_2-powder in the Anatase modification with an average particle size of 100 nm was supplied by KRONOS International INC (KRONOS 1002). The impurities are mainly P_2O_5, K_2O and sulfate. The active component V_2O_5 occurred in the Shcherbinaite modification (Gfe Environmental Technology Ltd). The specimens were homogenized by ball milling in cyclohexane for 1 h and subsequently dried in air at 100 °C. The annealing of the resulting powder was carried out in MgO-stabilized Zirconia (PSZ) crucible in an air chamber furnace (Nabertherm). The heating rate was 5 K/min and the final temperature (400 to 700 °C) was held for 4 h. For comparison, additional annealing cycles were done for 10 h to investigate the time behavior.

Phase Ananlysis

The phase content of the annealed powders has been investigated by X-ray-powder diffraction. A θ-θ-diffractometer with CuKα radiation was used (BRUKER AXS, D5005, variable divergence slits. position sensitive detector or scintillation counter). The phase analysis was carried out using the Diffrac-Plus /Search program. For higher precision of the lattice parameter, Guinier patterns (ENRAF NONIUS FR 552, quartz (1010) Johansson monochromator. AGFA Structurix D7 DW X-Ray film) were taken which enabled the use of monochromatic radiation and an internal Si-standard. The data were processed by the least square fit program PULVER [17] in order to gain lattice parameters.

Thermal Analysis and Morphology

Thermal analysis was needed to inspect weight loss and reaction temperatures. In order to avoid reactions a MgO-stabilized ZrO_2 crucible was used. For weight loss measurements, specimens were annealed in a chamber furnace (4 h and 10h). The transformation temperature for faster reactions was determined by a DTA run (Netzsch STA 429). The morphology of the catalyst materials was inspected by SEM (Philips XL 20) and an EDX detector for cation analysis (EDAX with UTW detector).

XPS Studies

The measurements were carried out using a modified LHS/SPECS EA200 MCD system equipped with a Mg K_α source (1253.6 eV, 168 W). Spectra were run with a pass energy of 48 eV. The binding energy scale of the system was calibrated using Au $4f_{7/2}$ = 84.0 eV and Cu $2p_{3/2}$ = 932.67 eV from foil samples. The powder samples were mounted on a stainless steel sample holder. The base pressure of the ultra-high vacuum (UHV) chamber was 1×10^{-10} mbar. The position of the sample holder in the analysis chamber can be well reproduced allowing a good comparison of absolute intensities of different samples. Charging of the samples was taken into account by using the identical O1s binding energy of 529.9 eV for TiO_2 and V_2O_5 [18]. To correct for charging, the O1s core level peak maxima of samples heat treated at 500 °C and 675 °C were shifted by 3.5 eV and 2.4 eV, respectively. A Shirley background was subtracted from all spectra before peak fitting with a 30/70 Gauss-Lorentz product function was performed. Atomic ratios were determined from the integral intensities of the signals, which were corrected using empirically derived sensitivity factors [19].

Catalytic Testing

The support TiO_2-powder KRONOS 1002 was mixed with 4 mole % V_2O_5 powder in a vibrating unit and heat treated at 500, 600 and 800 °C in air for 4 h. The heating rate was 5 K/min. During the heat treatment the V_2O_5 spread on the surface of the TiO_2. The oxidative dehydrogenation of propane to propene (ODP) was used as test reaction. Before testing, the catalyst material was uniaxially pressed at 10 to 20 MPa. For the catalytic experiments, a fixed-bed reactor (ø = 6 or 12 mm) made of quartz, operated at ambient pressure and equipped with on-line gas chromatography was used. A reaction mixture consisting of 40 Vol. % C_3H_8, 20 Vol. % O_2, 40 Vol. % N_2 was passed through the undiluted catalyst (0.2 - 1 g; d = 250 - 355 μm) packed between two layers of quartz of the same particle size. Total flow rates from 10 to 150 cm^3/min were used depending on the type of catalyst.

RESULTS AND DISCUSSION
X-Ray Phase Analysis

The samples can be identified in the following way: sample T99.5/ V0.5/ 400/ 4:

T99.5 = mole% of TiO_2/ V0.5 = mole% V_2O_5/ 400 = calcination temperature/ 4 = calcination time. Table 1 shows the starting composition and the qualitative phase content as found by XRD. The phase content was calculated from a simple analysis of the peak height using

$X_{phase} = I_{phase} / S I_{all\ phases}$

Since the V-content is comparably small this simplification does not produce too large errors.

Table 1 Weight loss and XRD-results of various sample compositions in the system $TiO_2 - V_2O_5$

Sample	Nr. in fig. 2, table 2	Weight loss [wt.%]	Anatase S-Q [wt.%]	Rutile S-Q [wt.%]	Shcherbinaite S-Q [wt.%]
T99.5/ V0.5/ 400/ 4		0.16	100	0	0
T99.5/ V0.5/ 450/ 4		0.15	100	0	0
T99.5/ V0.5/ 500/ 4		0.12	100	0	0
T99.5/ V0.5/ 500/ 10		0.36	100	0	0
T99.5/ V0.5/ 550/ 4		0.29	100	0	0
T99.5/ V0.5/ 600/ 4		0.27	100	0	0
T99/ V1/ 400/ 4		0.06	100	0	0
T99/ V1/ 450/ 4		0.12	100	0	0
T99/ V1/ 500/ 4		0.13	100	0	0
T99/ V1/ 500/ 10		0.39	98	0	2
T99/ V1/ 550/ 4		0.35	99	1	0
T99/ V1/ 600/ 4		0.25	95	5	0
T97/ V3/ 400/ 4		0.08	91	0	9
T97/ V3/ 450/ 4		0.12	91	0	9
T97/ V3/ 500/ 4		0.15	91	0	9
T97/ V3/ 500/ 10		0.45	89	2	9
T97/ V3/ 525/ 10	[3]	0.12	94	0	6
T97/ V3/ 550/ 4		0.21	90	3	7
T97/ V3/ 600/ 4		0.69	21	80 (ss)	0
T97/ V3/ 625/ 4		0.68	24	76 (ss)	0
T97/ V3/ 650/ 4		0.69	12	88 (ss)	0
T97/ V3/ 675/ 4		0.68	6	94 (ss)	0
T97/ V3/ 700/ 4		0.76	0	98 (ss)	2
T95/ V5/ 450/ 4		0.26	86	0	14
T95/ V5/ 500/ 4		0.27	82	0	19
T95/ V5/ 500/ 10		0.42	87	0	13
T95/ V5/ 525/ 10		0.11	91	0	9
T95/ V5/ 550/ 4		0.30	86	2 (ss)	12
T95/ V5/ 600/ 4		0.38	85	4 (ss)	11
T95/ V5/ 625/ 4		0.82	0	100 (ss)	0
T95/ V5/ 650/ 4		0.85	0	96 (ss)	4
T95/ V5/ 675/ 4	[4]	0.85	0	96 (ss)	4
T95/ V5/ 700/ 4		0.91	0	98 (ss)	2
T90/ V10/ 450/ 4		0.16	79	0	21
T90/ V10/ 500/ 4		0.22	86	0	14
T90/ V10/ 500/ 10		0.30	82	0	19
T90/ V10/ 525/ 10		0.14	84	0	16
T90/ V10/ 550/ 4		0.25	85	1 (ss)	14
T90/ V10/ 600/ 4		0.25	79	2 (ss)	20
T90/ V10/ 625/ 4		0.62	28	59 (ss)	14
T90/ V10/ 650/ 4		0.83	0	87 (ss)	13
T90/ V10/ 675/ 4		0.95	0	94 (ss)	6
T90/ V10/ 700/ 4		0.98	0	90 (ss)	10
T87.5/ V12.5/ 675/ 4		1.00	0	89 (ss)	11 (mod. Liquid)
T85/ V15/ 675/ 4		1.12	0	86 (ss)	15 (mod. Liquid)
T82.5/ V17.5/ 675/ 4		0.93	0	80 (ss)	20 (mod. Liquid)
T80/ V20/ 450/ 4		0.36	71	0	29
T80/ V20/ 500/ 4		0.44	55	0	45
T80/ V20/ 525/ 10		0.12	71	0	29
T80/ V20/ 550/ 4		0.05	54	2 (ss)	45 (mod. Liquid)
T80/ V20/ 600/ 4		0.20	60	2 (ss)	37 (mod. Liquid)
T80/ V20/ 625/ 4		0.51	24	44 (ss)	33 (mod. Liquid)
T80/ V20/ 650/ 4		0.68	0	69 (ss)	31 (mod. Liquid)
T80/ V20/ 675/ 4	[5]	0.56	0	78 (ss)	22 (mod. Liquid)
T80/ V20/ 700/ 4		0.86	0	77 (ss)	23 (mod. Liquid)
T50/ V50/ 675/ 4		0.46	0	49 (ss)	52 (mod. Liquid)
T5/ V95/ 675/ 4	[6]	0.39	0	0	100

Some of the phases showed variations with respect to peak position and height. Thus, in order to gain more accuracy, the spectra were measured by a Guinier camera. The derived lattice parameters are given in Table 2.

Table 2   Measured and theoretical crystal data

Material	Sample in Table 1	Phases	a_0	b_0	c_0
			[nm]	[nm]	[nm]
Reference Materials					
ICPDS 21-1272		Anatase	3.785	--	9.514
ICPDS 21-1276		Rutile	4.593	--	2.959
ICPDS 41-1426		Shcherbinaite	11.516	3.566	4.373
Starting Powders					
TiO_2	1	Anatase	3.785 ± 1		9.519 ± 2
TiO_2 annealed at 1200 °C	2	Rutile	4.593 ± 1		2.959 ± 1
V_2O_5	7	Shcherbinaite	11.519 ± 7	3.563 ± 2	4.370 ± 2
Catalyst Specimens					
97 mole% TiO_2 + 3 mole% V_2O_5	3	Anatase and	3.785 ± 1	--	9.515 ± 5
		Shcherbinaite	11.490 ± 1	3.564 ± 2	4.369 ± 2
95 mole% TiO_2 + 5 mole% V_2O_5	4	Rutile-(ss)	4.588 ± 1	--	2.958 ± 1
80 mole% TiO_2 + 20 mole% V_2O_5	5	Rutile-(ss) and	4.586 ± 1	--	2.958 ± 1
		Shcherbinaite	11.519 ± 1	3.565 ± 3	4.369 ± 4
5 mole% TiO_2 + 95 mole% V_2O_5	6	Shcherbinaite	11.527 ± 7	3.567 ± 2	4.362 ± 4

The phase relations are discussed for increasing temperatures:

25 °C – 525 °C: For temperatures up to 525 °C the mixed V_2O_5/TiO_2 pattern showed the reflexes of Anatase beside Shcherbinaite. Their lattice parameters are very close to the reference data (samples 1 and 2 in Table 2). There was no indication for further phases. Thus, one can conclude that no alloying has taken place. The contact between Anatase and Shcherbinaite could have only been formed during heat treatment most probably due to surface diffusion of the Vanadium species.

525 °C – 700 °C: For increasing temperatures, formation of Rutile was found starting at about 550 °C which is significantly lower than for pure TiO_2. The literature mentions some results for the formation of Rutile in the presence of various contaminants [23]. In this work the transformation, from Anatase to Rutile was encouraged by the presence of V_2O_5. In contrast, pure TiO_2 powders require temperatures of 900 to 1000 °C for the transformation the V_2O_5/TiO_2 catalysts transformed to the Rutile structure already at temperatures beneath 600 °C. Both the peak intensities are modified and the lattice parameters are shifted to smaller values (cf. Table 2). This indicates the formation of a Rutile solid solution (henceforth denoted as Rutile-(ss)) in

which the V_2O_5 is dissolved. In order to investigate the extension of the field of Rutile-(ss), a series of samples with compositions from 0 up to 20 mole% V_2O_5 has been annealed at 675 °C. The specimens with 3, 5 and 10 mole% V_2O_5 content exhibited only the reflexes of the Rutile-(ss). For 12.5 mole% and higher concentrations, an additional Shcherbinaite phase with modified peak positions was detected (cf. Table 2, sample 5). Thus the phase field of the Rutile-(ss) begins at 3-5 mole% V_2O_5 and 525 - 550 °C and extends to more than 10 mole% V_2O_5 at 675 °C.

The formation of a Rutile-(ss) may be regarded as a solid state reaction:

$$TiO_2 (Anatase) + V_2O_5 (Shcherbinaite) \Rightarrow Ti_{1-x} V_x O_y (Rutile-(ss))$$

This reaction was first observed under the conditions:

$$525 \text{ °C} < T < 550 \text{ °C for 3 mole% } V_2O_5 < x < 5 \text{ mole% } V_2O_5$$

This type of reaction could formally be a eutectoidal type reaction, which, however, would cause an endothermic signal on heating. A second possibility is a reaction from metastable starting phases to thermodynamically stable product phases which would cause a (sluggish) exothermic reaction. The DTA run on 10% V_2O_5 powder mixtures did not show any exothermic effect, the signal was very small, but the samples did react to Rutile-(ss) and changed their colour. Thus, simple eutectic reaction can be excluded. A Rutile-(ss) once formed at elevated temperatures could not be re-transformed to Anatase (annealing and XRD). This proves that the Rutile formation has to be seen as a reaction to form a thermodynamically stable product. The phase fields in the low temperature area have to be seen as metastable fields. The Rutile-(ss) field will extend down to room temperature in a stability phase diagram. For practical purposes, however, the metastable phase relations are of more importance.

Eutectic Reaction: For temperatures above 625 °C, Rutile-(ss) and a vanadium rich melt are co-existing. A eutectic melt was already mentioned in ref. [20], however, the publication presented only a melt projection showing a eutectic reaction in the V_2O_5-rich regime. Differential thermal analysis of a 5 mole % TiO_2- 95 mole % V_2O_5 specimen showed a hysteresis for heating and cooling cycles and a sharp peak for the eutectic temperature at 631 °C (cf. Fig. 1), which is taken as the eutectic temperature of this system.

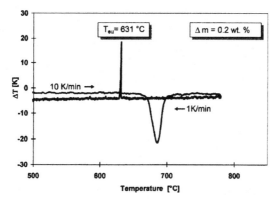

Figure 1   DTA-plot of a specimen with 95 mole % V_2O_5 and 5 mole % TiO_2

The V_2O_5-TiO_2-melt forms on heating. The DTA plot shows a sharp crystallization peak on cooling. The resulting TiO_2-containing V_2O_5-(ss) shows the crystallography of the Shcherbinaite, however, with different lattice parameters. This is a result of the alloying of Ti-ions into V_2O_5 on melting.

Preliminary Behaviour Diagram

Based on the XRD results a working behaviour diagram was deduced (Fig. 2). The numbers 1 to 7 are the numbers of specimens corresponding to those in table 2.

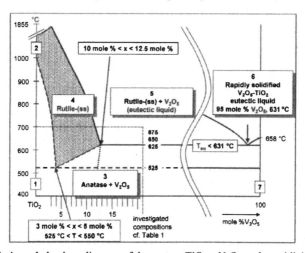

Figure 2 Preliminary behaviour diagram of the system $TiO_2 - V_2O_5$ under oxidizing conditions

The straight lines indicate the stable equilibria, while the dotted lines show the metastable diagram for the given conditions. A continuous melt regime, a eutectic reaction and an extended Rutile–(ss) belong to the stable part. The Anatase-Shcherbinaite miscibility field below 550°C can be considered as a metastable situation.

The unit cell of the Rutile-(ss) is smaller than that of the pure Rutile. Since V-ions both in the 4 and 5 valent state are smaller than Ti-ions, this behaviour indicates a substitutional replacement. For this Rutile-(ss) different defect models may be considered:

1) Vanadium remains entirely in the 5 valent state; no loss of oxygen occurs; charge compensation by interstitial oxygen ions:

$$V_2O_5 \xrightarrow{TiO_2} 2\,V_{Ti}{}^{\bullet} + O_i{}^{''} + 4\,O_O{}^x$$

2) Vanadium remains entirely in the 5 valent state; no loss of oxygen occurs; charge compensation by vacancies on the vanadium site:

$$2\,V_2O_5 \xrightarrow{TiO_2} 4\,V_{Ti}{}^{\bullet} + Vac_{Ti}{}^{''''} + 10\,O_O{}^x$$

3) Vanadium is reduced to the 4 valent state with loss of $\frac{1}{2}\,O_2$

$$V_2O_5 \xrightarrow{TiO_2} 2\,V_{Ti}{}^x + 4\,O_O{}^x + \tfrac{1}{2}\,O_2\,(v)\uparrow$$

This type of solid solution has been described by Bond [10], but those specimens have been annealed under reducing condition. Hence, a significant weight loss was reported which is due to the evaporation of oxygen (MS spectra). The major difference in this work is the use of the atmosphere during thermal treatment.

Because V_2O_5 is known to have high vapor pressure [14] the risk of evaporation had to be taken into account. If one assumes an evaporation of oxygen a significant weight loss should be observable. For the type (3) reaction the weight loss should be 1.8 wt.% for a 10 mole% and 3.2 wt.% for 20 mole% V_2O_5. However, the weight loss after annealing at different temperatures was ranging from 0.1 wt.% to 1 wt.% only (cf. table 1). There was no indication for an influence of the V_2O_5 concentration on the weight loss, but one for time and temperature. Additional ICP-OES investigations of the V-content showed that no significant change of the chemical composition could be recognized. Thus a type (3) defect model is unlikely for oxidizing conditions; V can not be entirely reduced to its 4 valent state. On the other hand, a change of the color of the samples was recognized. The V_2O_5/TiO_2 starting mixture is white or yellow whereas the reacted Rutile–(ss) is brown-grey. The darkness of the color increases with increasing V_2O_5 content and could be explained by a partly filled conduction band and metallic bonding character.

XPS studies were needed to determine the changes in oxidation state of V during temperature treatment. Figure 3 shows $V2p_{3/2}$ spectra of a sample with 10 mole% V_2O_5 (T90/V10/500/4 and T90/V10/675/4). The peak can be described as the sum of two components centered at 517.2 eV (A) and 515.9 eV (B). The position of component A agrees well with the $V2p_{3/2}$ binding energy we have observed for V_2O_5, in which vanadium is expected to be in its highest oxidation state

(+5). Therefore, we attribute component A to the V^{5+} oxidation state. As the *absolute* values of the binding energies of vanadium in all oxidation states vary significantly in the literature, the assignment of component B was based on the reported *differences* in the binding energies, ΔBE, of V^{5+} and lower oxidation states. Generally, the reported ΔBE between V^{5+} and V^{4+} is 0.7-1.0 eV, and the reported ΔBE between V^{5+} and V^{3+} is 1.2-1.5 eV [22-23]. This suggests that component B corresponds to V^{3+}.

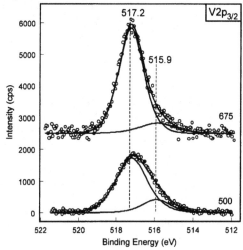

Figure 3 $V2p_{3/2}$ spectra of 10 mole % V_2O_5/90 mole% TiO_2 heat treated at various temperatures

Based on the spectral weight of the corresponding V^{5+} and V^{3+} states, the oxidation states of the sample T90/V10/500/4 and T90/V10/675/4 are calculated as 4.65±0.07 and 4.75±0.03, respectively. This means that both samples contain a mixture of vanadia in the 5 valent state and in the 3 valent state. The performed XPS analysis leads to the conclusion, that the oxidation state of the sample heat treated at 500 °C and 675 °C is the same for both samples within the experimental error.

Phase morphology
       At temperatures under 550 °C the initial phases Anatase and Shcherbinaite remain stable and the contact can be only formed via surface diffusion. A material prepared in this regime is characterized by fine and homogenous particles, cf. Fig. 4. This phase composition gives the best catalytic behavior of the system in the ODP (Table 3).

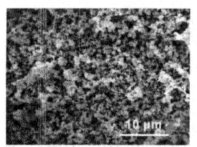

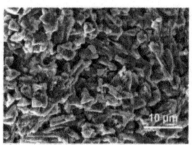

Figure 4 SEM-image of 5 mole% V_2O_5/
95 mole% TiO_2 annealed at 550 °C

Figure 5 SEM-image of 10 mole% V_2O_5/
90 mole% TiO_2 annealed at 675 °C

For temperatures between 550 °C and 631 °C an eutectoid solid state reaction takes place and Rutile-(ss) occurs. If a TiO_2 supported catalyst is being exposed to high temperatures, a substantial grain growth is observed; cf. Figure 5. Since the reaction starts from the two pure phases, the reaction path has to cross the stability field of the melt, which is expected to have much faster transport mechanism. The formation of a Rutile at temperatures higher than the eutectic temperature will proceed via a solution (V_2O_5-rich melt) and a re-precipitation (TiO_2-rich solid). The catalytic activity of this material is almost zero, cf. Table 3. For high V_2O_5-concentrations and temperatures above the eutectic temperature a liquid phase enhances the reaction rate, as can be seen in Fig. 6. Very large clearly facetted grains were formed. The EDX spectra show that these particles still contain the full amount of V_2O_5. The morphology is needle like which corresponds well with the orthorhombic structure of Shcherbinaite.

Figure 6 SEM-image of 95 mole% V_2O_5/5 mole% TiO_2 annealed at 675 °C

Catalytic Activity

The calcination temperature exerts a large influence on the catalytic activity in the ODP reaction (cf. Table 3).

Table 3   Catalytic results of the ODP at 500 °C of samples calcined at different temperatures under study (40 Vol. % C_3H_8, 20 Vol. % O_2, 40 Vol. % N_2, contact time $\tau = 0,75$ [$g \cdot s \cdot ml^{-1}$]) propane conversion (X), product selectivity (S) and propene yield (Y)

Catalyst	Calcination tempe-rature [°C]	X C_3H_8 [%]	Y C_3H_6 [%]	S C_3H_6 [%]	S CO [%]	S CO_2 [%]	S C_2H_6 [%]	S Oxyg [%]
KRONOS-TiO_2/V_2O_5	500	20.1	7.7	38.3	34.0	22.9	4.8	< 0.1
spreaded	600	18.4	7.2	39.0	35.6	22.6	2.8	< 0.1
	800	< 0.1	< 0.1	< 0.1	< 0.1	< 0.1	< 0.1	< 0.1

Only the material prepared at low temperatures (< 550 °C. cf. Fig. 4) exhibits acceptable catalytic behavior, i.e. these phases are catalytically active but metastable. As soon as the reaction to Rutile-(ss) occurs the catalytic activity collapses (cf. Fig. 5). The Shcherbinaite phase is no longer in contact to the gas phase and graingrowth is occured. Detailed studies on the catalytic activity and its correlation with morphology and phase content are depicted in [15].

CONCLUSION AND OUTLOOK
    The phase development in the system consisting of TiO_2 (Anatase) and V_2O_5 (Shcherbinaite) was investigated with TiO_2-V_2O_5 specimens prepared by ball milling and annealing in air from 400 to 700 °C. The XRD-results showed the formation of a Rutile-(ss). Investigation of the SEM-EDX and ICP-OES underlined that the Rutile being formed is a TiO_2-V_2O_5 solid solution.
A working behaviour diagram was deducted. For the temperature range up to 525 °C only Anatase beside Shcherbinaite is found for all compositions. This indicates a wide solubility gap. For higher temperatures, a metastable Rutile solid solution field extends up to about 12.5 mole% V_2O_5, in which the Vanadium is predomonantly in the 5 valent state. The reaction from Anatase and Shcherbinaite to the Rutile-(ss) begins at 3 mole% < V_2O_5 < 5 mole% and 525 °C < T < 550 °C. The reaction is considered as a sluggish reaction from metastable to stable phases.
For high V_2O_5 concentration an eutectic reaction was recognized at 631 °C. For higher temperatures the reaction from Anatase to Rutile might proceed via a liquid phase which results in a substantial growth of the initial 100 nm Anatase particles to form the final Rutile grains of 1 µm in size. V_2O_5/TiO_2 catalysts calcined at T > 600 °C show no catalytic activity since there is no accessible VO_x phase due to the solid solution of V_2O_5 in Rutile.

REFERENCES
[1]   G. Ertl, H.Knözinger, J.Weitkamp, "Preparation of Solid Catalysts", *Wiley-VCH*, (1999)
[2]   J.B. Stelzer, H. Kosslick, J. Caro, D. Habel, E. Feike, H. Schubert, „Aufbau und katalytische Aktivität hierarchisch strukturierter Oxid-Katalysatoren - Teil 1", *Chem. Ing. Tech.*,**7**, 872-77 (2003)
[3]   J.B. Stelzer, M.-M. Pohl, H. Kosslick, J. Caro, D. Habel, E. Feike, H. Schubert, „Aufbau und katalytische Aktivität hierarchisch strukturierter Oxid-Katalysatoren - Teil 2", *Chem. Ing. Tech.*, **11**, 1656-60 (2003)

[4]   R.J. Farranto, C.H. Bartholomew, "Fundamentals of Industrial Catalytic Processes", *Chapman & Hull*, (1997)

[5]   C.H. Bartholomew, R.T. Baker, D. Dadyburjor, "Stability of Supported Catalysts: Sintering and Redispersion", ed. J.A. Horsley, *Catalytic Studies Division*, (1991)

[6]   G.A. Fuentes, "Catalyst deactivation and steady-state activity: A generalized power-law equation model", *Appl. Catal.*, **15**, 33-40 (1985)

[7]   P.G. Menon, "Diagnosis of Industrial Catalyst Deactivation by Surface Characterization Techniques", *Chem. Rev.*, **94**, 1021-46 (1994)

[8]   M. Englisch, V.S. Ranade, J.A. Lercher, "Liquid phase hydrogenation of crotonaldehyde over Pt/SiO_2 catalysts", *Appl. Catal. A General*, **163**, 111-22 (1997)

[9]   M. Grzesik, J. Skrzypek, B.W. Wojciechowskie, "Time-on-stream catalyst decay behaviour in a fixed-bed catalytic reactor under the influence of intraparticle diffusion: intraparticle diffusion affects only catalytic reactions", *Chem. Eng. Science*, **47**, 4049-55 (1992)

[10] G.C. Bond, A.J. Sarkany, "The vanadium pentoxide-titanium dioxide system : Structural investigation and activity for the oxidation of butadiene", *J. Catal.*, **57**, 476-93 (1979)

[11] M. Piechotta, I.Ebert, J. Scheve, Strukturuntersuchungen an Mischoxidkatalysatoren. VI. TiO_2-V_2O_5", *Anorg. Allg. Chem.*, **368**, 10-17 (1969)

[12] G.C. Bond, J.P. Zurita, S. Flamerz, "Structure and reactivity of titania-supported oxides. Part 1: vanadium oxide on titania in the sub- and super-monolayer regions", *Appl. Catal.*, **22**, 361-78 (1986)

[13] G.C. Bond, S.F. Tahir, "Vanadium oxide monolayer catalysts Preparation, characterization and catalytic activity", *Appl. Catal.*, **71**, 1-31 (1991)

[14] A. Vejux, P. Courtine, "Interfacial reactions between V_2O_5 and TiO_2 (anatase): Role of the structural properties", *J. Solid State Chem.*, **23**, 93-103 (1978)

[15] J.B. Stelzer, A. Feldhoff, J. Caro, M. Fait, D. Habel, E. Feike, H. Schubert, „Aufbau und katalytische Aktivität hierarchisch nano-strukturierter Oxid-Katalysatoren - Teil 3", *Chem. Ing. Tech.*, **8**, 1086-92 (2004)

[16] R.D. Shanon, J.A. Pask, "Topotaxy in the anatase-rutile transformation", *Amer. Mineral.*, **49**, 1707-17 (1964)

[17] M.Weber, Institute for Applied Geosciences, TU Berlin, (1988)

[18] J. Kasperkiewics, J.A. Kovacich, D. Lichtman, "XPS studies of vanadium and vanadium oxides", *J. Electron Spectrosc. Relat. Phenom.*, **32**, 123-32 (1983)

[19] D.Briggs, M.P. Seah, "Practical Surface Analysis", *Wiley-VCH*, Chichester, (1990)

[20] S. Solacolu, M. Zaharescu, *Rev. Roum Chim.*, **17**, 1715-24 (1972)

[21] M.J. LaSalle, J.W. Cobble, "The Entropy and Structure of the Pervanadyl Ion", *J. Phys. Chem.*, **59**, 519-24 (1955)

[22] N.K. Nag, F.E. Massoth, "ESCA and gravimetric reduction studies on V/Al_2O_3 and V/SiO_2 catalysts", *J. Catal.*, **124**, 127-32 (1990)

[23] M.A. Eberhardt, A. Proctor, M. Houalla, D.M. Hercules, "Investigation of V Oxidation States in Reduced V/Al_2O_3Catalysts by XPS", *J. Catal.*, **160**, 27-34 (1996)

SYNTHESIS AND CHARACTERIZATION OF CUBIC SILICON CARBIDE (β-SiC) AND TRIGONAL SILICON NITRIDE (α-Si$_3$N$_4$) NANOWIRES

Karine Saulig-Wenger, Mikhael Bechelany, David Cornu*, Samuel Bernard, Fernand Chassagneux and Philippe Miele
Laboratoire des Multimatériaux et Interfaces UMR 5615 CNRS
University Claude Bernard - Lyon 1
43 Bd du 11 novembre 1918
F-69622 Villeurbanne Cedex, France
David.Cornu@univ-lyon1.fr

Thierry Epicier
GEMPPM UMR 5510 CNRS - INSA Lyon
20 Avenue Albert Einstein
F-69621 Villeurbanne Cedex, France

ABSTRACT

By varying the final heating temperature in the range 1050°C - 1300°C, cubic silicon carbide (β-SiC) and/or trigonal silicon nitride (α-Si$_3$N$_4$) nanowires (NWs) were prepared by direct thermal treatment under nitrogen, of commercial silicon powder and graphite. Long and highly curved β-SiC NWs were preferentially grown below 1200°C, while straight and short α-Si$_3$N$_4$ NWs were formed above 1300°C. Between these two temperatures, a mixture of both nanowires was obtained. The structure and chemical composition of these nanostructures have been investigated by SEM, HRTEM, EDX and EELS.

INTRODUCTION

Numerous studies have been recently devoted to ceramic and metallic nanowires (NWs) due to their outstanding properties which can be tailored by varying their chemical composition and also their crystalline structure. The possible applications of these NWs run from nanoelectronics to composite materials. Among the series, silicon-based NWs, made of cubic silicon carbide (β-SiC), silicon dioxide (SiO$_2$) or trigonal silicon nitride (α-Si$_3$N$_4$), are of particular interest due to their interesting mechanical[1,2], electrical[3,4] and/or optical[5] properties. However, industrial applications clearly need a cheap growth method for the large-scale fabrication of NWs. Moreover, works are also devoted to coaxial nanocables (NCs) due to their possible uses as reinforcement agents for mechanical applications, the outer sheet of the NC acting as an interface between the nanowire and the matrix.

Numerous routes have been reported for the fabrication of SiC NWs. The most promising ones for large scale production are (i) carbothermal reduction routes using carbon nanotubes (CNTs) as templates[6-8] or a mixture of SiO$_2$ nanoparticles and active carbon[9] and (ii) methods based on VLS (Vapour Liquid Solid) growth mechanism e.g. assisted by catalytic metallic nanoparticles[10-14]. Only few techniques have been described for the preparation of silicon nitride nanowires. They usually correspond to those reported for SiC NWs but they are driven at higher temperature (1400°C to 1650°C). The three main routes are the following: (i) carbothermal reduction of silica[15-19], (ii) the use of carbon nanotubes as templates[20] and (iii) a VLS growth technique[21]. All those synthetic methods require however, either expensive starting materials such as CNTs or expensive equipment.

We previously reported that the direct pyrolysis under nitrogen of commercial silicon powder in presence of graphite led to the formation of cubic silicon carbide (β-SiC) NWs[22,23]. These nanowires have diameters in the range 20 – 30 nm and micrometric lengths. We showed that when a mixture of argon and oxygen is used instead of nitrogen, amorphous silica nanowires were preferentially obtained[24,25]. As an extent to these results, we report in the present paper the influence of the final heating temperature on the structure and chemical composition of the resulting silicon-based nanostructures.

EXPERIMENTAL

All the experiments were conducted following the same experimental procedure. Silicon powder (Aldrich 99.999%, 60 mesh) was placed in an alumina boat containing a piece of graphite. This boat was then placed in the alumina tube of a horizontal tubular furnace, the tube being previously degassed *in vacuo* before filling with nitrogen (electronic grade). Under the gas flow (5 mL.min^{-1}), the boat was heated up to the selected temperature (heating rate 200°C min^{-1}), held for 1 hour then allowed to cool down to room temperature. Finally, powder was scraped from the alumina boat and analysed by scanning electron microscopy (SEM, Model N°S800, Hitachi), high-resolution transmission electron microscopy (HRTEM, Field Emission Gun microscope JEOL 2010F) and electron energy-loss spectroscopy (EELS, Digi-PEELS GATAN).

RESULTS AND DISCUSSION

In previous works, β-SiC nanowires have been obtained by heating silicon powder under nitrogen at 1200°C, this temperature being held during 1 hour before cooling down[22,23]. In order to examine the influence of the final heating temperature on the yield, the crystallographic structure and the chemical composition of the resulting nanowires, five independent experiments have been conducted up to 1050°C, 1150°C, 1200°C, 1250°C and 1300°C, respectively. In all these experiments, the final temperature was held during 1h before cooling down. Each crude product was first analysed by SEM and representative images of each sample are shown in fig.1.

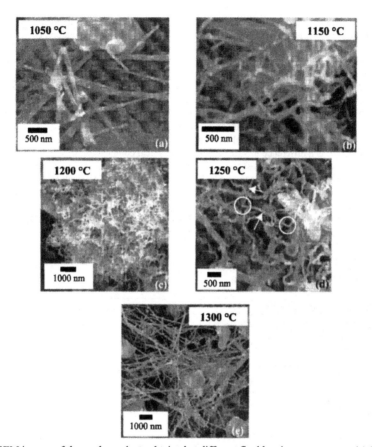

Fig. 1: SEM images of the crude products obtained at different final heating temperature: (a) 1050°C, (b) 1150°C, (c) 1200°C, (d) 1250°C and (e) 1300°C.

The crude sample heated up to 1050°C contained only few nanowires (Fig. 1a), located on the surface of the residual silicon particles. These NWs are straight with scattered diameters (from ~10 to ~300 nm) and short lengths (below ~3 μm). When the experiment was conducted up to 1150°C, it resulted in the formation of larger amount of nanowires (Fig. 1b). Their diameters are in the same range but they are longer with lengths estimated above ~8 μm. As illustrated by Fig. 1b, this result should be related with their highly curved shape. Nanowires exhibiting similar diameters and lengths were obtained at 1200°C (Fig. 1c). In that case, the yield, estimated from the SEM images, was however significantly improved. In contrast, when the thermal treatment was driven up to 1250°C, the yield was not significantly improved (Fig. 1d). At this temperature, two kinds of nanowires were observed: numerous long and highly curved nanowires similar to those obtained at 1200°C mixed with few straight nanowires (Fig. 1d, white arrows). The latter are shorter and exhibit well-defined angles which can be interpreted as changes in direction of the NW axis (Fig. 1d, white circles). The SEM image of the samples obtained after heating up to 1300°C shows a strong modification in the shape of the obtained NWs (Fig. 1e). No long and curved nanowires were detected but numerous straight NWs were observed. Their diameters are comparable to those previously obtained but their lengths are shorter and below ~8 μm.

HRTEM investigations coupled with EELS analysis have been conducted in order to determine if there is a difference in structure and/or chemical composition within the two kinds of NWs observed by SEM. Figure 2a shows a HRTEM image of a typical long and curved nanowire obtained at 1200°C. On the corresponding EELS spectrum (Fig. 2b), two main features are observed at ~100 eV and ~284 eV corresponding to Si-L and C-K edges, respectively. As expected, further selected area electron diffraction (SAED) analysis showed that cubic silicon carbide (β-SiC) nanowires have been formed. According to HRTEM investigation, a high density of stacking faults was observed (fig. 2a) which can be related to high growth rate, as previously mentioned[22,23].

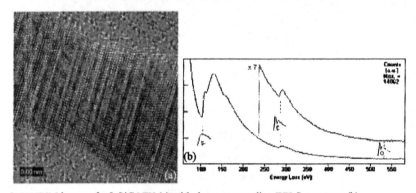

Fig. 2: HRTEM image of a β-SiC NW (a) with the corresponding EELS spectrum (b).

In contrast, figure 3a shows a typical HRTEM image of a straight and short nanowire obtained at 1300°C. The chemical composition of the observed NW has been determined by EELS analysis performed on its core (Fig. 3b). Two main features are observed at ~100 eV and ~400 eV corresponding to Si-L and N-K edges, respectively. Features at ~284 eV and ~532 eV, corresponding to C-K edge and O-K edges respectively, were not detected. A coating of ~2.5 nm thickness was observed around the analysed NW. Its EELS analysis performed using a 2 nm probe revealed that it is composed of carbon. As illustrated by figure 3c, fast Fourier transformation (FFT) shows that the core of the NW is composed of the trigonal polymorph of silicon nitride (α-Si₃N₄). Complementary structural analysis showed that the carbon coating is amorphous. The obtained nanostructures can be considered as carbon sheathed silicon nitride nanocables (α-Si$_3$N$_4$@C).

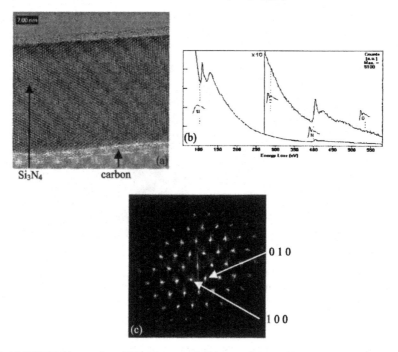

Fig. 3: (a) HRTEM image of a α-Si$_3$N$_4$ NW coated with amorphous carbon with corresponding (b) EELS spectrum and (c) fast Fourier transformation (FFT) of the core of the NW.

In previous works, we suggested that the silicon carbide nanowires were formed at 1200°C by a Vapour-Solid (VS) nucleation process, carbon being transported to silicon nanoparticles by nitrogen through the formation of CN-like derivatives[22,23]. For the formation of α-Si$_3$N$_4$ NWs *e.g.* at higher temperature (>1250°C), we can suggest a similar VS growth mechanism with a preferential nucleation of silicon nitride onto the surface of the silicon nanoparticles due to the higher temperature reached during the experiment. The formation of an amorphous coating of carbon onto the NWs can be interpreted by considering that carbon transportation by nitrogen was still effective at that temperature. Comparative analysis of nanowires obtained at 1300°C revealed that main α-Si$_3$N$_4$ NWs were free of carbon coating. Moreover, no SiC nanowires were detected in the sample obtained by heating up to 1300°C. This result is of primary importance for the determination of the growth mechanism of these nanowires because it indicates that their formation did not occurred when the temperature was increased, but either when the final temperature is reached or, less probably, during the cooling step.

While only α-Si₃N₄ NWs were formed at 1300°C, a mixture of β-SiC and α-Si₃N₄ NWs were obtained when the experiment was conducted up to 1250°C. This is evidenced by the bright field (Fig. 4a) and dark field (Fig. 4b) TEM images recorded on a whole mass of nanowires.

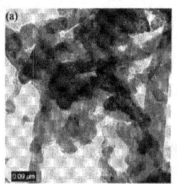

<u>Fig. 4</u>: bright field (a) and dark field (b) TEM images of a whole mass of β-SiC and α-Si₃N₄ NWs.

The image in bright field reveals the presence of Si₃N₄ nanowires, which appear in bright in Fig. 4a. The dark field image 4b, made with a β-SiC reflection, reveals that the Si₃N₄ NWs were mixed with SiC NWs which appear in bright in this image. These observations clearly point out that between 1200°C and 1300°C, a mixture of β-SiC NWs and α-Si₃N₄ NWs were obtained.

CONCLUSION
    The direct thermal treatment of a silicon powder under nitrogen yielded the formation of cubic silicon carbide (β-SiC) in the range 1050°C – 1200°C. As we found, when the experiment was driven in the range 1200°C – 1300°C, a mixture of β-SiC and trigonal silicon nitride (α-Si₃N₄) nanowires (NWs) were obtained. Moreover, when the experiment was conducted up to 1300°C, only α-Si₃N₄ NWs were observed within the sample. The growth mechanism for these nanowires were presumed to be a VS process, starting with the nucleation of silicon carbide or silicon nitride nuclei onto the surface of the silicon nanoparticles.

REFERENCES
[1]E.W. Wong, P.E. Sheehan, C.M. Lieber, "Nanobeam mechanics: elasticity, strength, and toughness of nanorods and nanotubes", *Science*, **277**, 1971-1975 (1997).
    [2]Y. Zhang, N. Wang, R. He, Q. Zhang, J. Zhu, Y. Yan, "Reversible bending of Si₃N₄ nanowire" *J. Mater. Res.*, **15**, 1048-1051. (2000)

[3]K. W. Wong, X. T. Zhou, F. C. K. Au, H. L. Lai, C. S. Lee, S. T. Lee, "Field-emission characteristics of SiC nanowires prepared by chemical-vapor deposition", *Appl. Phys. Lett.*, **75**, 2918-2920. (1999)

[4]Z. Pan, H.-L. Lai, F. C. K. Au, X. Duan, W. Zhou, W. Shi, N. Wang, C.-S. Lee, N.-B. Wong, S.-. Lee, S. Xie, "Oriented silicon carbide nanowires: synthesis and field emission properties", *Adv. Mater.*, **12**,1186-1190. (2000)

[5]D.P. Yu, L. Hang, Y. Ding, H.Z. Zhang, Z.G. Bai, J.J. Wang, Y.H. Zou, W. Qian, G.C. Xiong, S.Q. Feng, "Amorphous silica nanowires: intensive blue light emitters", *Appl. Phys. Lett.*, **73**, 3076-3078 (1998)

[6]W. Han, S. Fan, Q. Li, W. Liang, B. Gu, D. Yu, "Continuous synthesis and characterization of silicon carbide nanorods", *Chem. Phys. Lett.*, **265**, 374-378 (1997)

[7]C. C. Tang, S. S. Fan, H. Y. Dang, J. H. Zhao, C. Zhang, P. Li, Q. Gu, "Growth of SiC nanorods prepared by carbon nanotubes-confined reaction", *J. Cryst. Growth*, **210**, 595-599. (2000)

[8]J-M. Nhut, R. Vieira, L. Pesant, J-P. Tessonnier, N. Keller, G. Ehret, C. Pham-Huu, M. J. Ledoux, "Synthesis and catalytic uses of carbon and silicon carbide nanostructures", *Catal. Today*, **76**, 11-32. (2002)

[9]J. Pelous, M. Foret, R. Vacher, "Colloidal vs. polymeric aerogels: structure and vibrational modes", *J. Non-Cryst. Solids*, **145**, 63-70. (1992)

[10]X. T. Zhou, H. L. Lai, H. Y. Peng, F. C. K. Au, L. S. Liao, N. Wang, I. Bello, C. S. Lee, S. T. Lee, "Thin β-SiC nanorods and their field emission properties", *Chem. Phys. Lett.*, **318**, 58-62. (2000)

[11]B. Q. Wei, J. W. Ward, R. Vajtai, P. M. Azjayan, R. Ma, G. Ramanath, "Simultineous growth of silicon carbide nanorods and carbon nanotubes by chemical vapor deposition" *Chem. Phys. Lett.*, **354**, 264-268. (2002)

[12]X. T. Zhou, N. Wang, H. L. Lai, H. Y. Peng, I. Bello, N. B. Bello, N. B. Wong, C. S. Lee, S. T. Lee, "β-SiC nanorods synthesized by hot filament chemical vapor deposition" *Appl. Phys. Lett.*, **74**, 3942-3944. (1999)

[13]H. L. Lai, N. B. Wong, X. T. Zhou, H. Y. Peng, F. C. K. Au, N. Wang, I. Bello, C. S. Lee, S. T. Lee, X. F. Duan, "Straight a-SiC nanorods synthesized by using C-Si-SiO₂", *Appl. Phys. Lett.*, **76**, 294-296. (2000)

[14]Y. Zhang, N. L. Wang, R. He, X. Chen, J. Zhu, "Synthesis of SiC nanorods using floating catalyst", *Solid State Comm.*, **118**, 595-598. (2001)

[15]L. D. Zhang, G. W. Meng, F. Phillipp, "Synthesis and characterization of nanowires and nanocables", *Mater. Sci. and Eng.* A, **286**, 34-38 (2000)

[16]X. C. Wu, W. H. Song, W. D. Huang, M. H. Pu, B. Zhao, Y. P. Sun, J. J. Du, "Simultaneous growth of α-Si₃N₄ and β-SiC nanorods" *Mater. Res. Bull.*, **36**, 847-852 (2001)

[17]X. C. Wu, W. H. Song,B. Zhao, W. D. Huang, M. H. Pu, Y. P. Sun, J. J. Du, "Synthesis of coaxial nanowires of silicon nitride sheathed with silicon and silicon oxide", *Solid State Comm.*, **115**, 683-686 (2000)

[18]Y. H. Gao, Y. Bando, K. Kurashima, T. Sato, "Synthesis and microstructural analysis of Si₃N₄ nanorods", *Microsc. Microanal.*, **8**, 5-10. (2002)

[19]Y. H. Gao, Y. Bando, K. Kurashima, T. Sato, "Si₃N₄/SiC interface structure in SiC-nanocrystal-embedded α-Si₃N₄ nanorods", *J. Appl. Phys.*, **91**, 1515-1519. (2002)

[20]W. Han, S. Fan, Q. Li, B. Gu, X. Zhang, D. Yu, "Synthesis of silicon nitride nanorods using carbon nanotube as a template", *Appl. Phys. Lett.*, **71**, 2271-2273 (1997)

[21]H. Young Kim, J. Park, H. Yang, "Synthesis of silicon nitride nanowires directly from the silicon substrates", *Chem. Phys. Lett.*, **372**, 269-274. (2003)

[22]K. Saulig-Wenger, D. Cornu, F. Chassagneux, G. Ferro, T. Epicier, P. Miele, "Direct synthesis of β-SiC and h-BN coated β-SiC nanowires", *Solid State Comm.*, **124**, 157-161 (2002)

[23]K. Saulig-Wenger, D. Cornu, F. Chassagneux, G. Ferro, P. Miele, T. Epicier, "Synthesis and characterization of β-SiC nanowires and h-BN sheathed β-SiC nanocables", *Ceramic Transactions, (Ceramic Nanomaterials and Nanotechnology)* **137**, 93-99. (2003)

[24]K. Saulig-Wenger, D. Cornu, F. Chassagneux, T. Epicier, P. Miele, "Direct synthesis of amorphous silicon dioxide nanowires and helical self-assembled nanostructures derived therefrom", *J. Mater Chem.*, **13**, 3058-3061. (2003)

[25]K. Saulig-Wenger, D. Cornu, P. Miele, F. Chassagneux, S. Parola, T. Epicier, "Synthesis of Si-based nanowires", *Ceramic Transactions, (Ceramic Nanomaterials and Nanotechnology II)* **148**, 131-137. (2004)

HIGH ENERGY MILLING BEHAVIOR OF ALPHA SILICON CARBIDE

Maria Aparecida Pinheiro dos Santos
Instituto de Pesquisas da Marinha - IPqM , Grupo de Materiais
Rua Ipirú n° 2 - Ilha do Governador
CEP 21931-090, Rio de Janeiro/RJ, Brasil

Célio Albano da.Costa Neto
Programa de Engenharia Metalúrgica e de Materiais, COPPE – UFRJ,
CP 68505, CEP 21945-970, Rio de Janeiro/RJ, Brasil

ABSTRACT

Alpha silicon carbide (α- SiC) powder was comminuted in a planetary mill during the periods of time of ½, 2, 4 and 6h. The rotation speed was 300rpm, the milling media used was isopropilic alcohol and the grinding bodies were spheres of zirconia estabilized with ceria. The milling powders were characterized concerning the particle size distribution, chemical composition and density. It was observed a great reduction in the particle size, from micrometric to submicrometric size, showing the capability of producing even nanometric particles.

INTRODUCTION

Processing of powder in the nanometer range increases the sinterability of ceramic materials, especially for those that are covalent bonded. If the grain size is kept in nanometer range, promising functional and structural properties may arise. Also, small grain size and large amount of grain boundaries area can lead to superplasticity and extreme hardness, since hardness of nanophase ceramics often increases with decreasing grain size [1].

The majority of problems found in processing particles of submicron size arises from the high surface area created in the powders, which often outcome in surface contamination, low packing density and formation of agglomerates [2,3].

The present study has the objective of refining a coarse α-SiC powder produced in Brazil, not used for sintering, up to submicron size via planetary mill.

EXPERIMENTAL PROCEDURES

The starting material was α-SiC powder (Alcoa SiC 1000) with mean grain size of 5,5 µm, Figure 1, and purity of 98,7% [4]. This material was comminuted in a planetary mill (300 rpm) during ½, 2, 4h and 6h. The milling vessel was specific design for this study, having the capacity of 500 ml, made of stainless steel and coated with WC-CO by HVOF. The milling media was composed of isopropylic alcohol and ZrO_2 (Zirconox ®, Netsch Brasil Ltda) balls with size in the range of 0,7 to 1,2 mm.

Figure 1 - SEM photomicrograph of the as received SiC powder. The average particle size (d_{50}) was 1.77 μm and morphology showed marked sharp edges.

After milling, the powders were analyzed by particle size distribution (granulometer CILAS 1064), surface area (BET - Model Gemini III 2375), particle density (He picnometer -ACCUPYC 1330 Micromeritics). Before analyze the milled powders in the granulometer, they were ultrasonic dispersed for 5 minutes in a sodium pyrophosphate solution.

RESULTS AND DISCUSSION

Table 1 shows the average particle size (d_{50}) as a function of the milling time in isopropilic alcohol, and Table II shows the average particle size (d_{50}), surface area (S), the amount of ZrO_2 added to the system by the milling process and the powder density, either measured by picnometry and calculated by X-ray spectroscopy.

Table I- Average particle size (d_{50}) as a function of the milling time in isopropyl alcohol [4].

Milling time (h)	Average particle size d_{50} (μm)
Original (0)	1,77
½	0,41
2	0,38
4	0,36
6	0,36

Table II– Characteristics of α-SiC as received and milled for 4 and 6h [4].

t_m (h)	$d_{50}(\mu m)$	$S(m^2/g)$	ZrO_2 (%)	$d^*(g/cm^3)$ Picnom. He
-	1.77	3.483	-	3.269
4	<0,5	10.089	12	3.906
6	<0,5	11.479	16	3.968

It can be noted that the powder density has increased from a value of 3.2 g/cm³, measured for the as received SiC, to approximately 4 g/cm³ for 4 and 6 h of milling time. The increase in density was attributed to a major contamination of ZrO_2, as shown in Figure 2, which was detected by x-ray fluorescence. It is worth mention that it increased to 12% and 16 %wt of ZrO_2 for 4 and 6 hours of milling, respectively. The density of ZrO_2 is about 6 g/cm³, and the milled powder had its density increased form 3.2 to 3.9 g/cm³, an increase of 20%.

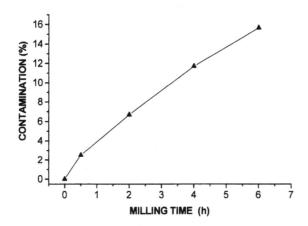

Figure 2 – The amount of zirconia, measured by X-ray fluorescence, added by the milling process as a function of the milling time.

Laser granulometry analysis showed that 100% of the particles have sizes bellow 1μm and the mean particle size was around 0.4 μm for the milling times of 4 and 6h. X-ray diffraction showed the presence of small broad peaks, and this type of feature is expected when particle size lies in the range of 0.1 to 0.5μm [5], which is confirmer in figure 3.

Figure 3 – High magnification of milled SiC powder showing the large amount of submicron particles, mainly SiC and also ZrO_2.

The present data showed that planetary mill can be used to produce submicron size powder in a very short time, since ½ h of milling resulted in mean particle size of 0.4 μm. The secondary additivation of the ZrO_2 was not observed to impart the sinterability of the material, since values of 99% density were obtained [4], and the fracture toughness was measured to be 5.8 MPa.m$^{1/2}$ [6]. This improves behavior was due to transformations of ZrO_2 into ZrC [4,7].

CONCLUSIONS

The planetary milling process was quite efficient in reducing the mean particle size of α-SiC ceramic. Techniques employed to evaluate particle size indicate that most of the particles lie in the nanometer range between 0.5 to 0.1 μm. Contamination of ZrO_2 was observed and it helped to density the material to 99% of the theoretical density. Further research has to be conducted in order to understand the effect of ZrO_2 during the sintering process.

REFERENCES

[1] H. Hahn, R. Averback , J.Am.Ceram.Soc. 74 (11), 2918 (1991).
[2] B.Matos, "Influência do meio na moagem ultrafina de carbeto de silício", Projeto de Formatura, Departamento de Engenharia Metalúrgica e de Materiais da Escola de Engenharia da Universidade Federal do Rio de Janeiro, Rio de Janeiro, Abril (2002).
[3] H.Tanaka , " Sintering of Silicon Carbide ", Silicon Carbide Ceramics –1 Fundamental and Solid Reaction – Edited by Shigeyuki Sõmiya and Yoshizo Inomata – Ceramic Research and Development in Japan Series – Elsevier Applied Science (1988).
[4] M.A. P. Santos, "Processamento e sinterização de Carbeto de Silício Nacional", Tese de Doutorado, PEMM, COPPE, UFRJ(2003).

[5] K. Jiang, " On The Applicability of The X-ray Diffraction Line Profile Analysis in Extracting Grain Size and Microstrain in Nanocrystalline Materials",Journal of Materials Research, V.14(2),pp.549-559 (1999).

[6] E.Souza Lima, "Mechanical Behavior of alfa - SiC Based Nanocomposites", II SBPMat

Rio de Janeiro, RJ (2003)

[7] C.K. Narula, Ceramic Precursor Technology and its Application,Macel Dekker.,New.

ACKNOWLEDGEMENTS

Thanks to Dr. Francisco Cristovão de Melo (IAE/CTA) and Dr. Leonardo Ajdelsztajn (UC Irvine) for helping this research.

SYNTHESIS OF BORON NITRIDE NANOTUBES FOR ENGINEERING APPLICATIONS

Janet Hurst
David Hull
NASA Glenn Research Center
21000 Brookpark Rd
Cleveland, Ohio 44135

Daniel Gorican
QSS Group
21000 Brookpark Rd
Cleveland, Ohio 44135

ABSTRACT
Boron nitride nanotubes (BNNT) are of significant interest to the scientific and technical communities for many of the same reasons that carbon nanotubes (CNT) have attracted wide attention. Both materials have potentially unique and important properties for structural and electronic applications. However of even more consequence than their similarities may be the complementary differences between carbon and boron nitride nanotubes. While BNNT possess a very high modulus similar to CNT, they also possess superior chemical and thermal stability. Additionally, BNNT have more uniform electronic properties, with a uniform band gap of 5.5 eV while CNT vary from semi-conductive to highly conductive behavior.

Boron nitride nanotubes have been synthesized both in the literature and at NASA Glenn Research Center, by a variety of methods such as chemical vapor deposition, arc discharge and reactive milling. Consistent large scale production of a reliable product has proven to be difficult. Progress in the reproducible synthesis of 1-2 gram sized batches of boron nitride nanotubes will be discussed as well as potential uses for this unique material.

INTRODUCTION

In the last decade, significant attention from the scientific community has been focused on the area of nanotechnology and specifically upon nanotube synthesis. While carbon nanotubes have generated the bulk of interest to date, other compositions offer promise as well and may have advantages or complementary properties relative to carbon nanotubes for various applications. At NASA Glenn Research Center, where application interests are often focused on high temperature propulsion, both BN and SiC nanotube synthesis are currently under investigation for high temperature structural and electronic materials (1,2). The focus of the current effort is BNNT synthesis. While boron nitride nanotubes are known to be structurally similar to carbon nanotubes, inasmuch as both are formed from graphene sheets, much less is known about BNNT. In large part this is due to the difficulty in synthesizing this material rather than lack of interest. It has been found that BNNT have excellent mechanical properties with a measured Young's modulus of 1.22 +/- 0.24 TPa (3). BNNT also have a constant band gap of about 5.5 eV (4). In contrast,

CNT vary from semi-conducting to conducting behavior depending on chirality and diameter of the product. Little progress has been demonstrated in the control of the chiral angle and hence the electronic properties of CNT. On the other hand, BNNT preferentially forms the zig-zag structure rather than the armchair or chiral structures due to the polar nature of the B-N bond (5). Recently, it has also been shown that BNNT systems have excellent piezoelectric properties, superior to those of piezoelectric polymers (6). Additionally, the expected oxidation resistance of BNNT relative to CNT suggests BNNT may be suitable for high temperature structural applications. This stability may be an important safety consideration for some applications, such as hydrogen storage, as carbon nanotubes readily burn in air. Many synthesis approaches have been tried with varying degrees of success. Among these approaches are pyrolysis over Co (7), CVD methods (8), arc discharge (9) laser ablation (10), and reactive milling techniques (11,12). The reported approach developed at NASA Glenn Research Center produced BNNT of significant length and abundance.

EXPERIMENTAL PROCEDURE

BNNT were prepared by reacting amorphous boron powder in a flowing atmosphere of nitrogen with a small amount of NH_3. Prior to heat treatment, fine iron catalyst particles were added in the range of up to several weight percent and briefly milled in polyethylene bottles with a hydrocarbon solvent and ceramic grinding media. Batch sizes of 2 grams are typically produced but the process should be easily scaleable to larger sizes. Milled material was applied to various high temperature substrates such as alumina, silicon carbide, platinum and molybdenum. Nanotubes were formed during heat treatments to temperatures ranging from 1100 C to 1400 C for brief times, 20 minutes to 2 hours.

Nanotubes were imaged with a Hitachi S4700 field emission scanning electron microscope with a super thin window EDAX Genesis System energy dispersive spectrometer (EDS) or Phillips CM200 transmission electron microscope operated at 200 kV, with Gatan electron energy loss spectrometer (EELS). Thermogravimetric analysis (TGA) was done in air up to 1000 C.

RESULTS

BNNT synthesized by this method are shown in Figure 1. The low magnification photo in Figure 1a shows an as-produced flake of BNNT. The flake was removed by tweezers from a 2.5 mm x 5 mm substrate or crucible. It is robust and easily handled in this form. Also, as the nanotubes are anchored within a growth media, there is no respiration hazard. BN nanotubes grow extensively both from the "top" and "bottom" of the growth media, where "top" refers to the side exposed to the atmosphere. Both top and bottom layers can be observed in Figure 1a as the edges were slightly rolled during handling. The nanotubes are quite long, 100 microns being common as shown in Figures 1b and 1c. On the bottom of the growth media, a layer of shorter nanotubes develops between the film and substrate or crucible. EDS also showed that the BNNT grow upon a film composed primarily of B, N, O, Fe and some purities. Nanotube diameters can be very consistent throughout their length; however this is processing temperature dependent. Not surprisingly, variation in the heat treatment temperature resulted in somewhat different products. A lower processing temperature resulted in more fine, uniform nanotubes, as those shown in figure 1b. Diameters of 20nm to 50 nm were typical. Higher temperatures and/or longer times resulted in large "nanotube" growth as shown in Figure 2a. Growth originated from fine nuclei but with excessive temperature, large structures quickly developed, up to a few microns in diameter. As shown in Figures 2a and 2b, secondary nucleation of small diameter BNNT also occurred on these larger structures. Higher processing temperatures, as well as excess catalyst concentration, resulted in interesting structures, such as nanohorns or nanoflowers, as seen in figure 2d. These structures also generally had considerable amounts of oxygen found by EDS, up to 6 w/o.

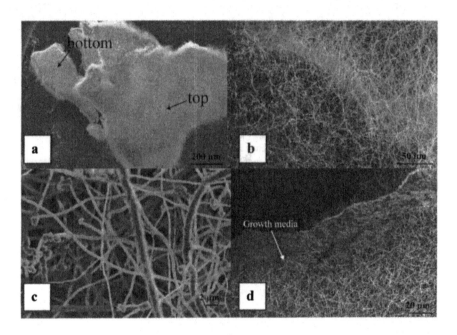

Figure 1 Field emission scanning electron microscope images of BNNT. (a) Typical flake peeled from substrate. (b)(c) Higher magnification photos of typical areas. (d) BNNT growing from media.

Transmission electron microscopy results are shown in Figures 3 and 4. TEM results showed the nanotubes to be nearly stoichiometric BN and a mixture of both straight walled nanotubes, as shown in Figure 3, and the "bamboo" structures, Figure 4. Predominately, the product from this method is multiwalled, commonly composed of 15-30 lattice layers, although this can be affected by processing conditions. Diameters of the

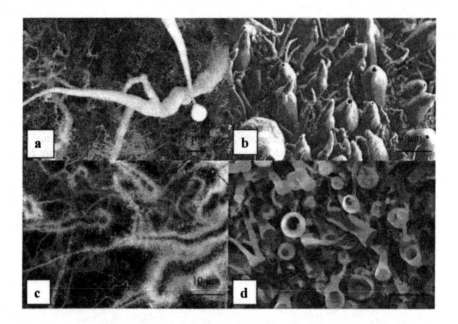

Figure 2. Examples of less typical structures synthesized at processing extremes.
a) Adjacent extremes in size. b) open ended nanopods c) fine BNNT nucleated on larger tubes
d) nanohorns.

Figure 3 TEM photos of typical straight walled BNNT

multiwalled BNNT were often in the 20-40 nm range, again determined by processing
conditions and catalysts concentration. Figure 3 shows the atomic planes within the
straight walled nanotubes exhibited lattice fringes at an angle of 12.5° with respect to the

tube axis. This has also been noted elsewhere (5, 13) and may be an indication of rhomohedral stacking order (12). In Figure 4 the typically highly faulted lattice walls of the bamboo structure are evident. These faulted short walls, with their open edge layers, have been suggested to be superior for hydrogen storage (14,15). The structure of the BN layers are analogous to stacked paper cups with potential hydrogen storage sites on the surface and between lattice planes, and also perhaps within the isolated voids. Predominately bamboo structures can be consistently produced by this processing method. Figure 5 shows a region of exclusively bamboo BNNT.

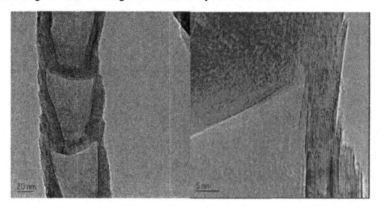

Figure 4 TEM photos of typical bamboo BNNT

The temperature stability of as-produced BN nanotubes was investigated by thermal gravimetric analysis. The results were compared to commercial as-produced carbon nanotubes*. Photos of the as-produced CNT and BNNT materials are shown in Figure 6 as well as those following heating in the TGA. The BNNT structure is clearly intact with the CNT decomposed, leaving behind the extensively oxidized iron catalyst. Figure 7 shows the TGA data in air, confirming that the carbon nanotubes have decomposed by 400 C. However, the BNNT are unaffected by the heat treatment with the exception of some slight weight gain from oxidation above 1000 C.

Figure 5.Regions of predominately bamboo nanotubes structures

Figure 6. As-processed BNNT and CNT samples were examined before and after exposure in air to 1000 C in a TGA. Photos on the left are as-processed samples. On the right are photos of material remaining following a TGA run to 1000C

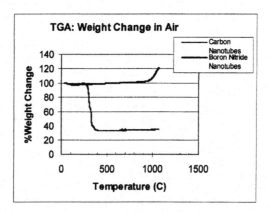

Figure 7. TGA results for BNNT and CNT samples showing the superior stability of BNNT relative to CNT

CONCLUSION

Boron nitride nanotubes were successfully and reproducibly grown by a NASA Glenn Research Center developed process. Currently 1-2 gram batches are being synthesized; however, the process is scaleable to much larger batch sizes. Sufficient amounts are now available so that boron nitride nanotubes are currently being incorporated into composites to provide high strength behavior at high temperature. It is also possible that this material may have many applications for sensors, electronics, piezoelectrics, among other applications of interest to NASA and the technical community.

A preponderance of bamboo structured nanotubes could be achieved by careful additions of catalyst materials and control of processing conditions. These bamboo structures are of interest for hydrogen storage applications. The boron nitride nanotubes were found to be much more stable at high temperatures than carbon nanotubes.

ACKNOWLEDGEMENTS

This work was sponsored by the Alternate Fuel Foundation Technologies (AFFT) Subproject of the Low Emissions Alternative Power (LEAP) Project at the NASA Glenn Research Center.

*Carbon Nanotechnologies, Inc.,16200 Park Row, Houston, TX

REFERENCES

[1] .J. Hurst NASA R&T 2002, NASA TM – 2002-211333.
[2] D. Larkin NASA R&T 2002, NASA TM – 2002-211333
[3] N.G. Chopra, A. Zettl, "Measurement of the Elastic Modulus of a Multi-wall Boron Nitride Nanotube", Solid State Commun. 105 (1998) 297.
[4] X. Blaise, A. Rubio, S.G. Louie and M.L. Cohen, Europhys. Lett. 28 (1994) 335.
[5] L. Bourgeois, Y. Bando and T. Sato, "Tubes of Rhombohedral Boron Nitride", J. Phys. D: Appl. Phys. 33 (2000)1902.
S. M. Nakmanson, A. Cazolari, V. Meunier, J. Bernholc, and M. Buongiorno Nardelli, "Spontaneous Polarization and Piezoelectricity in Boron Nitride Nanotubes", Phy. Rev. B 67, 235406 (2003).
[7] R. Sen, B.C. Satishkumar, A. Govindaraj, K.R. Harikumar, G. Raina, J.P. Zhang, A.K. Cheetham, C.N.R. Rao, "B-C-N, C-N and B-N Nanotubes Produced by the Pyrolysis of Precursor Molecules Over Co Catalysts", Chem. Phys. Lett. 287 (1998) 671.
[8] P. Ma, Y. Bando, and T. Santo, "CVD Synthesis of Boron Nitride Nanotubes Without Metal Catalysts", Chem. Phys. Lett. 337 (2001) 61-64.
[9] N.G. Chopra, R.J. Luyren, K. Cherry, V.H. Cresi, M.L. Cohen, S.G. Louie, A. Zettle, "Boron Nitride Nanotubes", Science, New Series, 269 (1995) 966.
[10] D.P. Yu, X.S. Sun, C.S. Lee, I. Bello, S. T. Lee, H. D. Gu, K. M. Leung, G.W. Zhou, Z. F. Dong, Z. Zhang, Appl. Phys. Lett 72 (1998) 1966.

[11.] Y. Chen, L.T. Chadderton, J.S Williams, and J. Fitz Gerald, "Solid State Formation of Carbon and Boron Nitride Nanotubes", Mater. Sci. Forum 343, 63 (2000).

[12.] Y. Chen, M. Conway, J.S Williams, J. Zou, "Large-Quantity Production of High-Yield Boron Nitride Nanotubes", J. Mater. Res. Vol 17 No.8 (2002) 1896-1899.

[13.] R. Ma, Y. Bando, T. Sato and K. Kurashima, " Growth, Morphology, and Structure of Boron Nitride Nanotubes", Chem. Mater. (2001) 13 2965.

[14.] R. Chen, Y. Bando, H. Zhu, T. Sato, C. Xu, and D. Wu, "Hydrogen Uptake in Boron Nitride Nanotubes at Room Temperature", J. Am. Chem. Soc. 124, 7672 (2002).

[15]. T.Oku, M. Kuno, I. Narita, "Hydrogen Storage in Boron Nitride Nanomaterials Studied by TG/DTA and Cluster Calculations", J. Phys. and Chem. of Solids, 65, 549 (2004).

COMPARISON OF ELECTROMAGNETIC SHIELDING IN GFR-NANO COMPOSITES

Woo-Kyun Jung
Seoul National University of Korea
301B/D 1255-1, San56-1, Shinlim, Kwanak,
Seoul, Korea, 151-742

Sung-Hoon Ahn
Seoul National University of Korea
301B/D 1205, San56-1, Shinlim, Kwanak,
Seoul, Korea, 151-742

Myong-Shik Won
Agency for Defence Development
Yuseong P.O.Box 35,
Daejeon, Korea, 305-600

ABSTRACT

The research on electromagnetic shielding has been advanced for military applications as well as for commercial products. Utilizing the reflective properties and absorptive properties of shielding material, the replied signal measured at the rear surface, or at the signal source can be minimized. The shielding effect was obtained from such materials that have high absorptive properties and structural characteristics, for example stacking sequence. In this research {glass fiber}/ {epoxy} / {nano particle} composites (referred to GFR-Nano composites) was fabricated using various nano particles, and their properties in electromagnetic shielding were compared. For visual observation of the nano composite materials, SEM(Scanning Electron Microscope) and TEM(Transmission Electron Microscope) were used. For measurement of electromagnetic shielding, HP8719ES S-parameter Vector Network Analyser System was used on the frequency range from 8 GHz to 12GHz. Among the nano particles, carbon black and Multi-walled Carbon Nano-tube (MWCNT) revealed outstanding electromagnetic shielding. Although silver nano particles (flake and powder) were expected to have effective electromagnetic shielding due to their excellent electric conductivities, test results showed relatively little shielding effect.

INTRODUCTION

Electromagnetic (EM) shielding is defined as the protection of the propagation of electric and magnetic waves from one region to another by using conducting or magnetic materials. The shielding can be achieved by minimizing replied signal or the signal passing through the material using reflective properties and absorptive properties of the material. This "Stealth" technology has been especially advanced for military application such as RADAR protection as well as for commercial products.

The shielding effect has been reported for such materials having high absorptive properties as Ferrite or Carbon-black [1-5]. Another mechanism of EM shielding was obtained from structural characteristics such as stacking sequence of composites [1, 6, 7]. Recently researchers have studied the electromagnetic properties of nano size particles [8-10].

In this research, Glass fiber reinforced composite made of {Glass fiber} / {Epoxy} / {Nano particles} were fabricated to observe EM shielding by nano particles mixed with Glass / Epoxy. Four types of nano particles were tested, and their properties in EM shielding were compared on X-band frequency (8GHz~12GHz).

EFFICIENCY OF ELECTROMAGNETIC SHIELDING

Figure 1 shows a schematic structure of a typical multi-layered material. A part of the incident EM wave is reflected at the surface while rest of the wave penetrates to inner materials. The wave undergoes surface reflection and interference at each layers to be absorbed, and finally attenuated wave escapes from the back of the material (transmitted wave).

EM-wave shielding is expressed by the shielding efficiency (SE), which uses unit of decibel (dB). The shielding efficiency is defined as the ratio of the power of incident EM-wave (P_I) to the power of transmitted EM-wave (P_T). The power of EM-wave can be related to electric field, incident electric field (E_I), and transmitted electric field (E_T) as Eq. (1) [11].

$$\text{SE(Shielding Efficiency)} = 10 \log (P_I / P_T)$$
$$= 20 \log |E_I / E_T| \qquad (1)$$

Appling the boundary conditions of transverse EM-wave, the relation among the electric fields of incident, reflected, and transmitted EM-wave of a multi-layered material is generally described as Eq. (2).

$$\begin{bmatrix} E_I \\ E_I{'} \end{bmatrix} = A_I^{-1} A_1 B_1^{-1} \cdots A_N B_N^{-1} \begin{bmatrix} 1 \\ k_T \end{bmatrix} E_T \qquad (2)$$

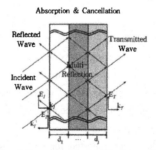

Figure 1. Schematic of electromagnetic shielding by a multilayer material.

From Figure 1, E_I' is the electric field of reflected electromagnetic-wave. Matrices A and B can be written as follows :

$$A_j = \begin{bmatrix} 1 & 1 \\ k_j & -k_j \end{bmatrix} \quad \text{and} \quad B_j = \begin{bmatrix} \exp(ik_j d_j) & \exp(-ik_j d_j) \\ k_j \exp(ik_j d_j) & -k_j \exp(-ik_j d_j) \end{bmatrix}$$

where, $i = \sqrt{-1}$, and k_j is the wave vector of the EM-wave in the j-th layer of multi-layered materials and d_j is thickness of the j-th layer. Minutely, k_j is the function of permittivity (ε_j), permeability (μ_j), and conductivity (σ_j) of the j-th layer at a given frequency ($f = \omega/2\pi$) such as

$$k_j^2 = \omega^2 \mu_j \varepsilon_j + i\omega\mu_j\sigma_j .$$

The EM-wave shielding obtained from Eqs. (1) and (2) is described as sum of the contributions due to reflection (SE_R), absorption (SE_A), and multi-reflections (SE_M) in the following Eqs. (3)~(6).

$$SE = SE_R + SE_A + SE_M \quad \text{(dB)} \tag{3}$$

$$SE_R = 20\log\left|\frac{(1+n)^2}{4n}\right| \quad \text{(dB)} \tag{4}$$

$$SE_A = 20\,\mathrm{Im}(k)d\log e \quad \text{(dB)} \tag{5}$$

$$SE_M = 20\log\left|1 - \frac{(1-n)^2}{(1+n)^2}\exp(2ikd)\right| \quad \text{(dB)} \tag{6}$$

where, n is the index of refraction of the shielding material, and $\mathrm{Im}(k)$ is the imaginary part of the wave vector in the shielding material.

According to the analysis of S parameter of two-port network system, transmittance (T), reflectance (R), and absorbance (A) can be described as :

$$T = |E_T / E_I|^2 = |S_{12}|^2 \ (=|S_{21}|^2) \tag{7}$$

$$R = |E_R / E_I|^2 = |S_{11}|^2 \ (=|S_{22}|^2) \tag{8}$$

$$A = 1 - R - T \tag{9}$$

From Eqs. (7)~(9), the effective absorbance (A_{eff}) can be described as

$$A_{eff} = (1-R-T) / (1-R)$$

with respect to the power of the effectively incident EM-wave inside the shielding material. It can be expressed as the form such as Eqs. (10) and (11).

$$SE_R = -10 \log (1 - R) \quad \text{(dB)} \tag{10}$$

$$SE_A = -10 \log (1 - A)$$

$$= -10 \log (T / (1 - R)) \quad \text{(dB)} \tag{11}$$

This analysis of the contribution of absorption and reflection to total EM-wave shielding is highly relevant for practical applications of the shielding materials.

EXPERIMENTAL

Specimen fabrication

Four types of nano particles (Table 1) were mixed with epoxy resin (YBD-500A80, *Br* series, Kukdo chemical Co.) and plain weave glass fabric (KN 1800, 300g/m^2, KPI Co.). The mixture of nano particles and epoxy resin were agitated by homogenizer to have 5wt% of particles in the mixture. After two hours of agitation, the mixture impregnated glass fabric. The impregnated three-phase composite was heated up to 120□ and held for one minute to become a prepreg. The GFR-nano composite parts were then fabricated by cutting into desired geometries, laying up, and curing these prepregs.

Specimens to measure EM-wave shielding were fabricated by curing process shown in Figure 2. The prepreg material was cured in a hot plate for two hours at 120⊓ and 5.5 atm.

In order to see the effect of stacking sequence, each specimen was made either in cross-ply or in quasi-isotropic laminate. To see the effect of thickness on EM-wave shielding, 8 plies and 16 plies were laid for each type of specimens

For the visual observation of the nano composites and nano particles, Scanning Electron Microscope (SEM, JSM-5600) and Transmission Electron Microscope (TEM, JEM-3000F) were used.

Table 1. Nano particles used in this study

Nano Particle	Average Diameter (nm)
Multi-Walled Carbon Nano-Tube(MWCNT)	15
Carbon Black(CB)	40
Silver Nano Flake(SNF)	40
Silver Nano Powder (SNP)	40

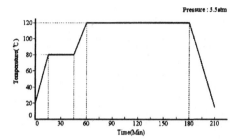

Figure 2. Cure cycle of GFR-nano composites.

Measurement of electromagnetic-wave shielding

The measurement of EM-wave shielding was performed using S parameter Vector Network Analyzer System (HP8719ES). The frequency range from 8GHz to 12GHz was scanned. From the measured data of reflected signal S_{11} and transmitted signal S_{21}, transmission coefficient and reflection coefficient were obtained. Using these coefficients, EM-wave shielding by absorption and reflection were calculated. Figure 3 shows the test setup for EM-wave shielding measurement.

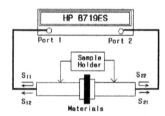

Figure 3. Test setup for electromagnetic shielding measurement.

RESULTS AND DISCUSSION

Morphology

Figure 4 shows TEM pictures of nano particles used in this study. The nano particles had 15nm⎤40nm average diameter. MWCNT showed 15nm average diameter and aspect ratio over 1,000.

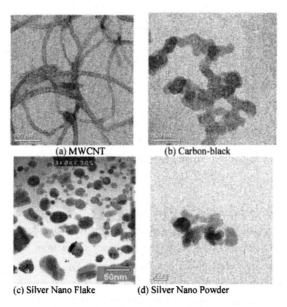

Figure 4. TEM pictures of nano particles used in this study.

The SNF had thin shell shapes while SNP had deformed ellipsoidal shapes.

Using SEM, structures of GFR-nano composites was observed. Although the cured epoxy and glass fiber of the composite material were observed with magnitudes between 1,000 and 2,000, the mingled nano particles, however, was not observable. Figure 5 (a) shows surface of the composite after grinding along fiber direction with 1,000 magnification. Figure 5 (b) is the cross-sectional view of the GFR-nano composite with 2,000 magnification. Even with higher magnification of 10,000, only the MWCNT was visually observed separately from the epoxy base (Figure 5 (c)). Lumped with epoxy, CB particles were not distinguishable, nor SNF and SNP particles.

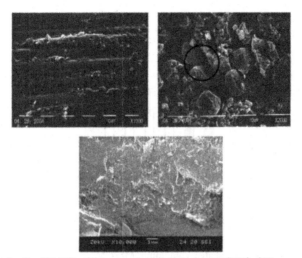

(a) Along fiber direction (X1,000)   (b) Cross section (X2,000)   (c) MWCNT in GFR- nano composites
(X10,000)

Figure 5. SEM pictures of GFR-nano composites.

Evaluation of electromagnetic-wave shielding

The measured EM-shielding efficiency was evaluated by characteristics of nano particles, thickness of specimens, and directions of fiber.

Figure 6 is an experimental data (GFR-MWCNT, 16C) of EM shielding as a function of frequency ranged from 8GHz to 12GHz.

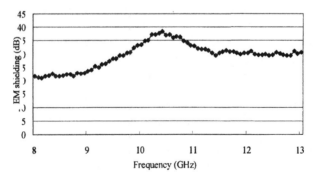

Figure 6. EM shielding of GFR-MWCNT (16C) composites.

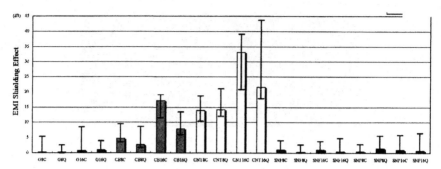

Figure 7. Comparison of EM shielding : Each nano particles, Stacking sequence, Thickness

Figure 7 shows the results of EM shielding of GFR-nano composites with various parameters. The raw GFR composites (G8C, G8Q, G16C, G16Q) had low EM shielding effect while GFR-nano composites containing MWCNT and CB (CB and CNT series) showed excellent shielding efficiency. Especially, GFR-MWCNT composites were three times higher efficiencies than those of CB series. The silver composites (SNF and SNP series) had higher efficiency than raw GFR-composites. The specimens of 8 ply laminate were 1.30mm thick in average, and specimens of 16 ply laminate were 2.50mm thick in average. Except for SNF and SNP series, as composite becomes thicker, shielding efficiency was increased. However, silver composites showed no change in shielding efficiency.

To investigate the effect of fiber orientation, prepergs were laid up with two conditions. One was cross-ply [0_8], and the other was quasi-isotropic [$(0 / 45)_2$]$_S$. In case of raw GFR composites and GFR-CB composites, cross-ply lay-up showed superior shielding efficiency to quasi-isotropic lay-up. For thin (8 plies) GFR-MWCNT composites, cross-ply lay-up had higher shielding efficiency than quasi-isotropic lay-up. Thick (16 plies) GFR-MWCNT composites showed different distribution as a function of frequency range. At lower frequency range (under 10GHz), cross-ply lay-up shielded EM-wave more efficiently. At higher frequency range (over 10GHz), to the contrary, quasi-isotropic lay-up had higner shielding efficiency. The SNF and SNP series had no special effects on the shielding efficiency.

CONCLUSIONS

Through a series of experiments, EM-wave shielding of GFR nano composites was observed. Four nano particles added in GFR composites were investigated by measuring EM-wave shielding of the composite materials. CB and MWCNT showed outstanding shielding effect of 17dB~43dB on 10GHz frequency range, but silver nano particles did not show significant effect. Besides the electromagnetic properties, the addition of nano particles may improve the structural properties of composite, for which further study is required.

REFERENCES

[1]J. H. Oh, K. S. Oh, C. G. Kim, and C. S. Hong, "Design of radar absorbing structures using Glass/epoxy composite containing carbon black in X-band frequency ranges," *Composites Part B*, **35**, 49 - 56 (2004).

[2]M. S. Pinho, M. L. Gregori, R. C. R. Nunes and B. G. Soares, "Performance of radar absorbing materials by waveguide measurements for X and Ku-band frequencies," *European Polymer Journal*, **38**, 2321 - 2327 (2002).

[3]D. D. L. Chung, "Electromagnetic interference shielding effectiveness of carbon materials," *Carbon*, **39**, 279 - 285 (2001).

[4]G. Li, G. G. Hu, H. D. Zhou, X. J. Fan and X. G. Li, "Absorption of microwaves in $La_{1-x}Sr_xMnO_3$ manganese powders over a wide bandwidth," *Journal of Applied Physics*, **90**, 5512 - 5514 (2001).

[5]K. B. Cheng, S. Ramakrishna and K. C. Lee, "Electro magnetic shielding effectiveness of copper/glass fiber knitted fabric reinforced poly-propylene composites," *Composites Part A*, **31**, 1039 - 1045 (2000).

[6]S. A. Tretyakov and S. I. Maslovski, "Thin Absorbing Structure for all incidence angles based on the use of a high-impedance surface," *Microwave and Optical Technology Letters*, **38**, 175 - 178 (2003).

[7]K. Matous and G. J. Dvorak, "Optimization of Electromagnetic Absorption in Laminated Composite Plates," *Transactions on Magnetics*, **39**, 1827 - 1835 (2003).

[8]A. Kajima, R. Nakayama, T. Fujii and M. Inoue, "Variation of dielectric permeability by applying magnetic field in nano-composite Bi_2O_3 - Fe_2O_3 - $PbTiO_3$ sputtered films," *Journal of magnetism and magnetic materials*, **258**, 597 - 599 (2003).

[9]L. I. Trakhtenberg, E. Axelrod, G. N. Gerasimov, E. V. Nikolaeva and E. I. Smirnova, "New nano-composite metal-polymer materials : dielectric behaviour," *Non-Crystalline Solids*, **305**, 190 - 196 (2002).

[10]P. Talbot, A. M. Konn and C. Brosseau, "Electromagnetic characterization of fine-scale particulate composite materials," *Journal of magnetism and magnetic materials*, **249**, 481 - 485 (2002).

[11]Y. K. Hong, C. Y. Lee, C. K. Jeong, D. E. Lee, K. Kim and J. Joo, "Method and apparatus to measure electromagnetic interference shielding efficiency and its shielding characteristics in broadband frequency ranges," *Review of Scientific Instruments*, **74** , 1098 - 1102 (2003).

DENSIFICATION BEHAVIOR OF ZIRCONIA CERAMICS SINTERED USING HIGH-FREQUENCY MICROWAVES

M. Wolff, G. Falk, R. Clasen
Saarland University, Department of Powder Technology
Building 43, D-66123 Saarbrücken, Germany

G. Link, S. Takayama, M. Thumm
Forschungszentrum Karlsruhe GmbH, IHM
Hermann-von-Helmholtz-Platz 1, D-76344 Eggenstein-Leopoldshafen, Germany

ABSTRACT
    The aim of the investigation was the formation of zirconia ceramics with a fine-grained microstructure and low porosity. The potential of high frequency (30 GHz) microwave sintering in this field was studied. The application of microwaves allowed high heating rates and short processing times because of volumetric heating and enhanced sintering kinetics and thereby a better control of the microstructure. Green bodies were made of tetragonal (8 mol-% Y_2O_3) and cubic (10 mol-% Y_2O_3) nanoscale and submicron zirconia powders, dispersed in aqueous suspensions. They were prepared by electrophoretic deposition (EPD) and had relatively high green densities and a homogenous pore size distribution.
    Green bodies and ceramics were characterized by means of XRD, SEM, density measurements by means of the Archimedes method and Hg-porosimetry. The main focus of this work is on the investigation of grain size evolution as a function of density.

INTRODUCTION
    In this work, the application of advanced green body shaping techniques, in combination with sintering using high-frequency microwaves was investigated to determine the possibilities of grain size and density control. The microstructure of ceramic materials has an important effect on their mechanical, electrical and optical properties. Porosity and grain size affect especially hardness, optical transmittance, thermal and ionic conductivity. The production of highly dense ceramics with small grain sizes makes high demands on shaping as well as on sintering methods. The use of fine powders as raw materials is required to obtain fine-grained microstructures in ceramics. But shaping of homogeneous ceramic green bodies with high densities, in particular, from submicron or nanosized powders, is a challenge. Dense and homogenous green bodies with small pores are a precondition for homogenous densification and low porosity after sintering. Removing large pores during sintering is much more difficult than elimination of small ones, sometimes even impossible and requires higher temperatures, which may result in strong grain growth. Furthermore, nanosized ceramics show slow grain growth only up to a certain critical temperature [1]. Thus substantial increase of density without grain growth cannot be achieved with long dwelling times at high temperatures.
    While the primary particle size of the powders is determined by their physical characteristics, the degree of particle agglomeration and the density of the green bodies depend on the shaping process. Dry pressing methods are not suitable to shape nanosized and submicron powders, because the compaction of very fine particles is difficult, due to their high degree of aggregation. Dry pressing of such powders results in non-homogenous pore size distributions and mechanical stresses within the green bodies. Electrophoretic deposition (EPD) from aqueous sus-

pensions is a simple and fast method to produce homogeneous and highly dense structures and was used to shape the green bodies for the investigations in this work [2]. For optimal shaping with EPD a deaggregation of the powders is required. The dispersion of the powders into stable primary particles and thus the supply of an agglomerate-free suspension, as well as high solids loadings, are fundamental conditions to produce green bodies with high green densities.

The heating process for sintering is an important step in the processing of engineering ceramics. Different sintering techniques such as conventional sintering (CS), microwave sintering (MWS), spark plasma sintering, sinter forging, hot pressing and hot isostatic pressing are applied in order to control the microstructure. The use of microwaves allows transfer of energy directly into the volume of dielectric materials. Through absorption mechanisms such as ionic conduction, dipole relaxation and photon–phonon interaction, the energy is converted to heat. Using MWS, volumetric heating can be realized, which allows the application of high heating rates and significantly shortens the processing time. Furthermore, the densification process of ceramic bodies seems to be enhanced by sintering in a microwave field, allowing a reduction of sintering temperatures and/or dwelling times compared to the processing in a standard sintering furnace [3, 4].

Heating of ceramics under microwave radiation depends on the dielectric properties. The absorbed power P inside the material is proportional to frequency ω and dielectric loss ε'', which itself is a function not only of frequency but also of temperature T (Eq. 1). This loss factor increases with temperature as well as with frequency in the microwave range for many low-loss ceramics.

$$P = \frac{1}{2}\omega\varepsilon_0\varepsilon''(\omega,T)|E|^2 \qquad (\text{Eq. 1})$$

Even though microwaves with a frequency of 2.45 GHz and 915 MHz have the broadest application, high power, millimeter-wave sources (gyrotrons) are an interesting alternative [5, 6]. In the case that the installed microwave power is not sufficient to reach sintering temperatures for heating experiments with low-loss ceramics, it is possible to preheat the ceramic by infrared radiation to temperatures where coupling of the ceramic samples is sufficiently high for microwave heating. Another possibility is to change to heating with higher frequencies such as 30 GHz, since power absorption is by a factor of 10–100 higher compared to the absorption at magnetron frequencies.

The sintering behavior of tetragonal stabilized zirconia (TZP) in a conventional and a microwave furnace with heating rates of 12 K/min was investigated in [7]. Maximal densities of 99.5 %TD were obtained at 1250 °C in the MW furnace whereas same densities were determined at 1350 °C in a conventional furnace. Important differences in the densification level were noticed between 1100 and 1300 °C. This was attributed to accelerated diffusion especially of O^{2-}-ions. No significant differences in grain structure were determined. An investigation in "ultrafast" sintering with 100 K/min in the MW furnace resulted in slightly lower sintered densities and crack generation.

Nightingale [8] compared the density-grain size relationship of conventionally and microwave sintered TZP and cubic stabilized zirconia (CSZ). The activation energy for grain growth is significantly lower in cubic than in tetragonal zirconia, which results in much stronger grain growth of cubic zirconia. For microwave sintered TZP samples, significantly smaller grain sizes were found at densities < 96 %TD and a slightly smaller grain sizes at higher densities, suggesting that microwaves tend to accelerate lattice diffusion more than they do surface and grain boundary diffusion. Nevertheless no differences were determined for CSZ, indicating a balance between the different diffusion processes. Accelerated grain growth after ageing at 1500 °C for 15 h was determined in [4], comparing MWS and conventional sintering. The grain size of the TZP samples was 1.5 times greater for MWS than for conventionally sintered ones.

EXPERIMENTAL

The green bodies were manufactured from nanosized powders of 8 mol-% yttria stabilized zirconia NA8Y (8Y Zirconium Oxide, Nanostructured and Amorphous Materials, USA) and powders of 10 mol-% yttria fully stabilized zirconia (TZ10Y, Tosoh Corp., Japan). The powders were characterized by XRD and TEM. The NA8Y powder had a mean grain size of 50 nm and consisted of monoclinic and tetragonal phase. TZ10Y had a mean aggregate size of 0.5 μm and consisted of cubic phase.

Electrophoretic deposition (EPD) was applied to manufacture the green bodies. As preparation for EPD, the powders were dispersed in water and the suspensions were electrosterically stabilized by tetramethylammoniumhydroxide (TMAH) and a dispersing aid. The dispersion process was carried out by means of a Dispermat N1–SIP (VMA-Getzmann GmbH, Germany), using a serrated disc. Electrophoretic deposition (EPD) was carried out in a horizontal cell (10 x 40 x 40 mm) under constant voltage, applying the membrane process [9]. The green bodies were first dried at room temperature for 24 h, then in a drying chamber at 120 °C for 24 h. Both drying processes were performed in air. The pore sizes were determined by Hg-porosimetry (Pascal 440, CE Instruments, Italy). The green bodies made of NA8Y had a green density of approximately 56 % of the theoretical density (TD) and those of TZ10Y, 54 %TD. The pore size distribution of the green bodies was homogenous and monomodal. NA8Y and TZ10Y had a mean pore size of 12 nm and 60 nm, respectively.

The conventional sintering experiments were carried out in air at 1300-1700 °C. The microwave sintering experiments were performed in a compact gyrotron system operating at 30 GHz in air with sintering temperatures starting from 1300 °C up to 1600 °C and a heating rate of 50 K/min. To avoid the evolution of an inverse temperature profile, which is a specific feature of microwave volumetric heating, the samples were sintered within a ZrO_2 crucible, placed in a box of mullite ceramic fiber boards, for thermal insulation.

Densities of the sintered samples was measured using the Archimedes principle, assuming a theoretical density of 6.07 g/cm^3 for tetragonal Y-ZrO_2 and 5.95 g/cm^3 for cubic Y-ZrO_2. To determine the grain sizes, the samples were thermally etched at 1240 °C for 30 min. Characterization of grain structure was performed by high resolution scanning electron microscopy (HRSEM). Grain sizes were measured applying the linear intercept technique, using a4i software (Aquinto AG, Germany).

RESULTS AND DISCUSSION

The conventional sintering experiments were carried out at a heating rate of 10 K/min, while microwave sintering (MWS) was conducted at 50 K/min. After heating, the samples were cooled down directly without a dwelling time. Thermal insulation and thus cooling behavior of the conventional furnace was different from the gyrotron system. The gyrotron system allowed much higher cooling rates than did the conventional furnace, since the thermal mass of the sample and thermal insulation was much smaller

In Fig. 1, the sintering behavior of NA8Y at different sintering temperatures is shown. At 1300 °C, CS resulted in a higher density than did MWS. This may be attributed to the lower heat-

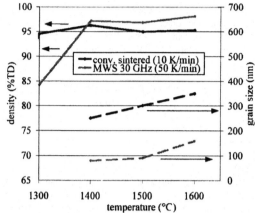

Fig.1: Density and estimated grain sizes after sintering of NA8Y
(tetragonal ZrO₂) in a conventional furnace and in a gyrotron system

ing rate in the case of CS. For higher sintering temperatures, densities of the microwave sintered samples were higher than for conventionally sintered ones, resulting in 1 %TD higher densities at 1400 °C, 2 % at 1500 °C and 3 % at 1600 °C.

Furthermore, the evolution of the grain sizes depended on the sintering techniques. MWS resulted in smaller grain sizes. Already at 1400 °C, in an earlier sintering stage, the conventionally sintered samples clearly had larger grain sizes. This is a possible explanation for the higher densification reached with MWS at higher temperatures. As the grain sizes remained smaller, a better elimination of porosity and thus a more homogenous densification resulted in higher densities.

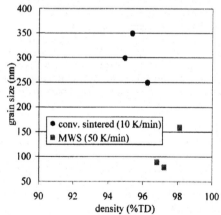

Fig. 2: Grain size as function of density after sintering of NA8Y
(tetragonal ZrO₂) in a conventional furnace and in a gyrotron system

Fig. 3: Microstructure of NA8Y: a) sintered with microwaves in the gyrotron system (50 K/min),
b) conventionally sintered (10 K/min) at 1500 °C

It is remarkable that even at higher sinter temperatures and clearly higher sintered densities, respectively, the grain size of the microwave sintered, tetragonal ZrO_2 remained smaller than
the conventionally sintered ones. In Fig. 2, the grain sizes are presented as function of the sintered density. At a density of around 95 %TD, the conventionally sintered samples had grain
sizes of about 300 nm. The maximum grain sizes of the MWS samples were about 150 nm, having a sintered density of 98 %TD. For sintered densities of 97 %TD, the grains were smaller than
100 nm. SEM pictures of the fracture surface of the samples, sintered at 1500 °C, are presented in
Fig. 3 a) and 3 b). As it can be seen, the morphology of the conventionally sintered samples was
clearly coarser than that of the microwave sintered samples.

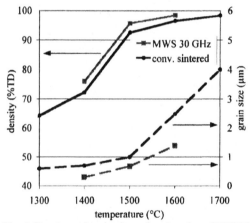

Fig. 4: Density and grain sizes after sintering of TZ10Y
(cubic ZrO_2) in a conventional furnace and in a gyrotron system

The sintering behavior of cubic zirconia (TZ10Y) is presented in Fig. 4. Densities and
grain sizes after heating up to 1600 °C of the samples are shown. The samples were cooled down
directly after heating. A difference in sintering behavior for conventional and microwave sinter-

ing can be determined for densification as well as for grain sizes. At 1400 °C, conventionally sintered samples had a density of 73 %TD whereas sintering with the gyrotron system results in densities of 76 %TD. Similar differences in density can be observed after sintering at 1500 °C (92 and 95 %TD) and 1600 °C (96 and 98 %TD). Simultaneously, MWS resulted in smaller grain sizes than CS, even though they had a higher sintered density. At 1400 °C, the average grain size of the MWS samples was about 300 nm whilst the conventionally sintered ceramics had grain sizes of about 700 nm.

A difference in grain growth can also be observed after sintering at 1500 °C. The average grain sizes of conventionally sintered samples reach 1 μm and 680 nm after sintering with microwaves. A stronger difference is determined after sintering at 1600 °C. The microstructure of the conventionally sintered samples was clearly coarser than the microwave sintered ones, having grain sizes of 2.5 and 1.4 μm, respectively.

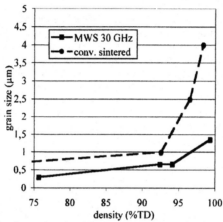

Fig. 5: Grain size as function of sintered density of TZ10Y (cubic ZrO_2), sintered in a conventional furnace and in a gyrotron system

The interaction of densification and grain growth is presented in Fig. 5. where grain size is shown as function of density. It can be observed, that at comparable sintered densities, conventional sintering resulted in explicitly coarser morphology. As can be seen in Fig. 4, in the conventional furnace, it was necessary to sinter at higher temperatures to reach the same densities as in the gyrotron system. At those higher sintering temperatures, strong grain growth occurred. Thus with a density of 92 %TD, the microwave sintered samples had average grain sizes of 600 nm. In order to reach comparable densities to conventional sintering, the samples already had grain sizes of 1 μm. Strong grain growth occurred mainly for densities higher than 92 %TD. Having densities of 97 %TD, conventionally sintered TZ10Y had grain sizes of 4 μm. Note that using CS, temperatures of 1700 °C were necessary to reach those densities. At a density of 98 %TD, the microwave sintered samples still had grain sizes < 1.5 μm. SEM images of polished and thermally etched samples of microwave and conventionally sintered TZ10Y are shown in Fig. 6 a) and 6 b). Distinct differences in morphology can be observed.

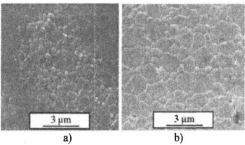

a)   b)

Fig.6: SEM images of TZ10Y, sintered a) in a gyrotron system (density 95 %TD, average grain size 700 nm) and b) in a conventional furnace (92 %TD, 1 μm) at 1500 °C

For a reliable comparison of process temperatures from different sintering techniques as presented in this work, the different heating mechanisms between conventional and microwave sintering and the resulting differences in temperature distribution within the ceramic body must be considered. In standard sintering furnaces, while heating, temperature gradients are induced in ceramics due to the low penetration depth of infrared radiation. This results in thermal gradients with a hot surface and a colder interior, which is compensated by a thermal conduction process depending on the thermal conductivity of the sintered material. The thermal conductivity of zirconia is low: tetragonal stabilized zirconia as well as cubic stabilized zirconia exhibit $\lambda \sim 2$ $Wm^{-1}K^{-1}$, which is even much smaller in a powder compact. Therefore, it is necessary to apply an optimized time-temperature program with relatively low heating rates to avoid crack formation or even destruction of the samples due to thermal stresses.

Using MWS, a volumetric heating can be realized which allows the application of high heating rates. On the other hand, a reverse problem to the conventional heating, namely temperature gradients from the hotter inside of the ceramic body to the surface, can exist. Within the gyrotron system, this effect is avoided by placing the samples within a zirconia crucible surrounded by mullite fiber boards for thermal insulation.

Furthermore, measurement of temperature was different for the sintering methods presented which may lead to systematic errors and therefore to wrong conclusions if the measured temperature is compared from different systems. The thermocouple of conventional furnaces usually is placed on the thermal insulation of the furnace wall. Thus the temperature is not detected directly next to the sample but it is assumed that the furnace has a homogenous temperature distribution. In the gyrotron system, the thermocouple is in contact with the sample surface. Nevertheless, especially for high heating rates, higher temperatures inside the sample may occur.

As exact comparison of temperature is difficult for the different sintering methods, the relationship between density and grain size shows reliable differences in densification behavior. Here the temperature parameter is not needed to discuss microwave specific effects on sintering. In Fig. 2 and Fig. 5 it can clearly be seen that the different sintering methods resulted in different sintering behavior. Using microwave sintering, higher sintered densities could be reached and the grains were significantly smaller than after conventional sintering. Furthermore it was possible to densify the ceramic with slower grain growth. As the grains remained small at an advanced sintering stage, higher maximum final sintered densities could be reached. The advanced sintering behavior can be attributed to higher sintering rates and an enhanced densification behavior induced by the microwave field.

CONCLUSION

The application of high frequency microwave sintering techniques in comparison with conventional sintering on tetragonal and cubic zirconia was investigated. A combination of advanced shaping and advanced sintering methods was carried out. The green bodies were formed by means of electrophoretic deposition (EPD) to produce homogenous and dense structures. Sintering experiments were carried out in a compact gyrotron system at 30 GHz. For high frequency microwave and conventional sintering, different densification behavior was found. As microwave sintering allows higher heating rates and enhanced densification, smaller grain sizes could be achieved for tetragonal as well as for cubic zirconia during intermediate and final sintering stage. Furthermore, higher final sintered densities were reached by applying microwave sintering. This is related to the smaller grain sizes during densification and thus a better elimination of porosity. Distinct differences in grain growth in relation to the densification state of the zirconia ceramics were observed. At their maximum sintered densities, microwave sintered tetragonal zirconia had nearly half as big grain sizes as conventional sintered ones. An even stronger effect was detected for cubic zirconia. Conventional sintering resulted in 2.5 times coarser grains.

ACKNOWLEDGEMENTS

The financial support from the Deutsche Forschungsgemeinschaft (DFG, Graduiertenkolleg 232) and the Helmholtz-Gemeinschaft (VH-FZ-024) is gratefully acknowledged.

REFERENCES
[1]   M. J. Mayo, D. C. Hague and D.-J. Chen, "Processing Nanocrystalline Ceramics for Applications in Superplasticity," *Mater. Sci. Eng. A,* **166** 145-159 (1993).
[2]   J. Tabellion and R. Clasen, "Electrophoretic Deposition from Aqueous Suspensions for Near-shape Manufacturing of Advanced Ceramics and Glasses - Applications," *J. Mater. Sci.,* **39** 803-811 (2004).
[3]   M. A. Janney, H. D. Kimrey, W. R. Allen and J. O. Kiggans, "Enhanced Diffusion in Sapphire During Microwave Heating," *J. Mater. Sci.,* **32** 1347-1355 (1997).
[4]   S. A. Nightingale, D. P. Dunne and H. K. Worner, "Sintering and Grain Growth of 3 mol% Yttria Zirconia in a Microwave Field," *J. Mater. Sci.,* **31** 5039-5043 (1996).
[5]   H. D. Kimrey, M. A. Janney and P. F. Becker, "Techniques for Ceramic Sintering Using Microwave Energy"; pp. 136-137 in *Int. Conference on Infrared and Millimeter Waves,* ed. Edited by Orlando, 1987.
[6]   G. Link, L. Feher, M. Thumm, H. J. Ritzhaupt-Kleissl, R. Böhme and A. Weisenburger, "Sintering of Advanced Ceramics Using a 30-GHz, 10-kW, CW Industrial Gyrotron," *IEEE Trans. Plasma Sci.,* **27** [2] 547-555 (1999).
[7]   A. Goldstein, N. Travitzky, A. Singurindy and M. Kravchik, "Direct Microwave Sintering of Yttria-stabilized Zirconia at 2.45 GHz," *J. Europ. Ceram. Soc.,* **19** 2067-2072 (1999).
[8]   S. A. Nightingale, H. K. Worner and D. P. Dunne, "Microstructural Development during the Microwave Sintering of Yttria-Zirconia Ceramics," *J. Am. Ceram. Soc.,* **80** [2] 394-400 (1997).
[9]   R. Clasen, S. Janes, C. Oswald and D. Ranker, "Electrophoretic Deposition of Nanosized Ceramic Powders"; pp. 481-486 in *Ceram. Transactions,* ed. Edited by H. Hausner, G. L. Messing and S. Hirano. Am. Ceram. Soc., Westerville (USA), 1995.

MANUFACTURING OF DOPED GLASSES USING REACTIVE ELECTROPHORETIC DEPOSITION (REPD)

Dirk Jung, Jan Tabellion and Rolf Clasen
Saarland University, Department of Powder Technology
Building. 43, D-66123 Saarbrucken, Germany

ABSTRACT

Doped glasses can be manufactured by means of gas infiltration, soaking of green bodies, by using powder mixtures or by melting. The melting point of silica glass is 2100 °C and most dopants evaporate at temperatures in this range. Because the sintering temperature of silica glass is about 700 °C lower then melting point, dopants, that would evaporate during the melting process, can be used. But it is difficult to achieve homogeneous green bodies by using powder mixtures because separation occurs as the particles have different sizes and densities. In case of the soaking method, during the drying process a surface segregation of the salt ions leads to samples with an inhomogeneous distribution. A promising method to manufacture homogeneous silica green bodies is the electrophoretic deposition (EPD) [1]. In the first approximation the mobility of the particles is independent of their size [2].

A modification of the EPD, by adding salts to the suspension, leads to reactive electrophoretic deposition (REPD). By varying the amount of added salts to the suspension, the dissolved ions modify the occupancy of particle surfaces and the composition of the cloud of ions. The adsorbed ions can be co-deposited with the particles leading to a very homogeneously doped green body. It is tested first for a suspension of SiO_2 that contained different amounts of boric acid and cobalt chloride. It is shown that the green bodies doped with boric acid can be sintered at lower temperatures compared to undoped ones. However, the sintering temperature depends on the amount of boric acid added to the suspension before.

INDRODUCTION

By doping of silica glass with different types and amounts of additives important properties like thermal expansion coefficient, refractive index or color can be tailored. Conventional doped glasses are fabricated via melting-routes. The melting point of silica glass is 2100 °C. In this case most of dopants cannot be used because they evaporate at 2100 °C.

Nano-sized, fumed silica powders (DEGUSSA OX50, A380) can be sintered at 1300 °C [3] because of their large sintering activity. Due to the sintering temperature being lower than the melting point, dopants can be used that normally evaporate at 2100 °C. Using powders to manufacture glasses offers alternative methods like gas infiltration, soaking of green bodies and employing powder mixtures to dope them. Gas infiltration is limited by the small variety of possible species and by the pore size of the green body. The pore size has to be larger than the mean free length of path of the gas molecules. In the case of green body soaking, ion-movement to the green body surface during the drying process leads to inhomogeneously-doped glasses [4]. Using powder mixtures to fabricate green bodies the different particle densities and sizes lead to decomposition and inhomogeneous green bodies are achieved. Another suspension-based method to fabricate green bodies using powders is electrophoretic deposition (EPD) [5]. In the case of aqueous slips the decomposition of water at applied DC>1.5 V leads to gas bubble formation at the electrodes, which are assembled into the green body. Using an ion-permeable membrane [6], which divides the electrophoresis cell into a suspension chamber and a second chamber, the deposition and recombination of ions are separated in space. Figure 1 illustrates the principle of the membrane method.

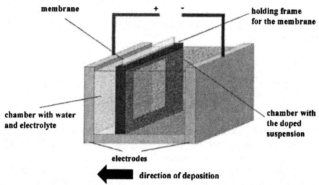

Figure 1: Schematic of the electrophoretic deposition by membrane method

Via EPD near-netshape manufacturing is possible since the surface tension of suspensions of silica particles ($\approx$ 20-70 mN/m) is lower than that of silica glass melts (340 mN/m) [7]. Moreover, it is possible to shape at room temperature. Via EPD it is possible to fabricate silica green bodies with a relative density of 45 %. In ref. [8] silica green bodies were fabricated by means of EPD from aqueous suspensions with a bimodal mixture of powders with a mean particle-size of 15 μm (SE15) and 50 nm (OX50). Maximum green density of 84 %TD was reached by a SE15 to OX50 ratio of 10:90. At the same time, the linear shrinkage was lowered to 4.7 %.

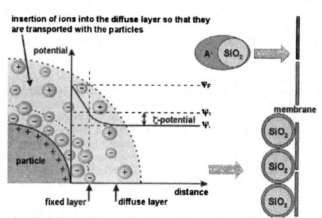

Figure 2: Functional principle of the reactive electrophoretric deposition (REPD)

Due to the dissociation of water there are ions in suspension. These ions arrange around the dispersed particles to compensate their surface charge. They form a cloud, which consists of anions and cations [9]. By shaping for EPD, a part of these ions will be co-deposited with the particles. The difference of reactive electrophoretic deposition (REPD) is the addition of water-soluble salt to the suspension to modify the cloud of ions. In Figure 2, the functional principle is schematically shown.

EXPERIMENTAL

Boric acid and cobalt chloride were used as a model system to test if REPD works with anions and cations. The experiments were carried out with a fused silica powder, which has a mean particle-size of 40 nm (DEGUSSA OX50) and a broad particle size distribution. Furthermore, boric acid (Roth) with a purity of 99.9 % and cobalt chloride (Fluka) with a purity of 98 % were used as dopants. The pH-value was adjusted using tetramethylamoniumhydroxide (TMAH), which is a strong base. The suspensions were prepared by dispersing nanosized OX50 in bidistilled water by means of a dissolver (PC-Laborsystem, Typ LDV1). Boric acid (anion doping) and cobalt chloride (cation doping) respectively were added to the suspension and the pH-value was adjusted to 11 by using TMAH.

The doped silica green bodies were shaped by REPD under constant electrical field by means of the membrane method [6]. The second chamber was filled with bidistilled water containing different amounts of TMAH. The conductivity was adjusted by adding TMAH to the 5-, 10- and 20-fold of the suspension's conductivity. In a modified experiment, boric acid or cobalt chloride were added to the liquid in the second chamber, to analyze if these ions pass the membrane to be deposited in the green body. After shaping, the green bodies were dried at room temperature for 12 hours. The content of ions was measured by inductively coupled plasma emission spectrometry (ICP-ES) (Jobin Yvon, JY 24). Furthermore, the green bodies were sintered at 900, 1000, 1100, 1200, 1300, 1350 and 1400 °C in a zone-sintering furnace.

The optical properties of the sintered samples were studied with an UV/VIS-spectrometer (Brucker, IFS 66v). Sintering kinetics were measured with a vertical dilatometer (Linseis, L75).

RESULTS AND DISCUSSION

The variation of the sintering temperature is an appropriate method to determine qualitatively the amount of boron oxide in the green bodies, because the concentration of boron oxide decreases the sintering temperature. In Figure 3, the linear shrinkage behavior of different doped green bodies is shown. The deposition was carried out from suspensions containing 0, 1, 5, 10 and 15 wt.% of boric acid. It is obvious, that the sintering temperature decreased with an increasing amount of boric acid in the suspension. Sintering of green bodies without boron oxide started at a temperatures of 1100 °C. 15 wt.% of boric acid decreased the first sintering stage to 900 °C. The zeta-potential of silicon dioxide at pH 11 was about −70mV. Due to the negative zeta-potential it was expected that only cations can co-deposit, because a negative zeta potential results in adsorption of cations. This is reasonable since a different electrical charge leads to an attractive force. Boron acid generates the anion $B(OH)_4^-$. Because of the negative zeta-potential it is remarkable that doping of green bodies via electrophoretic deposition using boron acid is successful. A co-deposition of anions and particles proves that anions as well as cations are part of the cloud of ions around the particles. To determine the incorporated amount of ions in the green body, an ICP analysis was performed to determine the final quantity of elements so that the efficiency could be calculated.

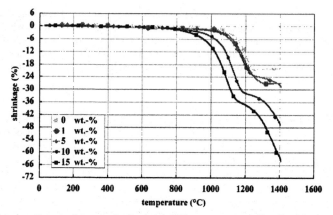

Figure 3: Sintering diagram for green bodies with different contents of boric acid

In Figure 4, the content of boron oxide in the green body as a function of the amount of boric acid in the suspension is shown. There is a linear correlation between the quantity of ions in the suspension and that in the green body. As shown in the diagram, less than half of the quantity in the suspension can be determined in the green bodies. As a matter of fact, only ions are co-deposited that are in the cloud around the particles. The quantity of ions in green bodies has to be lower than in suspension. Moreover, it is shown that an addition of boric acid in the second chamber also increases the concentration of boron oxide in the green bodies. Hereby, the suspension contained 5 wt.% and the water in the second chamber 0, 1, 3 and 5 wt.% of boric acid. In this case the amount of boric acid in green bodies was enhanced from 1.75 wt.% (0 wt.% in second chamber) to 3.6 wt.% (3 wt.% in second chamber).

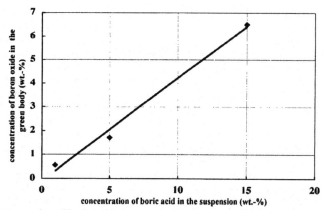

Figure 4: Concentration of boric oxide in green bodies increases with increasing amount of boric acid in suspension

Due to the negative zeta-potential a co-deposition of cations and particles is conceivable, due to the adsorption of cations needed to compensate the surface charge of particles. In Figure 5, the correlation between the concentration of cobalt chloride in the suspension and that in the green body is shown. The quantity of cobalt chloride in the green bodies is a linear function of cobalt chloride content in the suspension. Obviously, a higher deposition efficiency compared to the use of boric acid can be reached.

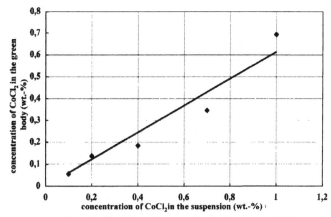

Figure 5: Concentration of cobalt chloride in the green bodies increases with increasing amount of cobalt chloride in suspension

Adding cobalt chloride to the liquid of the second chamber resulted in green bodies with an inhomogeneous structure can be observed in a gradient of color. Deposition proceeded from the cathode to anode. As the second chamber is almost saturated with the cations, the movement of those ions from suspension to this chamber is low. Anions from the second chamber can pass the membrane and displace the cations out of the green body, resulting in a gradient In Figure 6, two sintered glasses doped with cobalt chloride are shown.

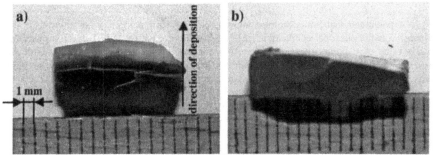

Figure 6: a) Sintered sample, deposited with cobalt chloride in both chambers of the electrophoresis cell. b) cobalt chloride only in the suspension chamber

The optical micrograph on the left hand side shows a glass that is produced with cobalt chloride in the second chamber. Decomposition can be clearly noted. This effect is not visible in the picture on the right-hand side. This glass was fabricated using cobalt chloride in only the suspension chamber. Boric acid-doped green bodies could also have a gradient in the concentration of ions but it could not be analyzed. Analyzing via ICP-ES needs dissolved samples using HF. That is why one is not able to observe gradients in the distribution of ions. The measurement of a gradient by means of sintering kinetics is also impossible because the total shrinkage was measured.

Furthermore, the transparency of the cobalt colored glasses was analyzed. Green bodies have been manufactured using suspensions with 0, 0.1, 0.2, 0.4, 0.7 and 1 wt.% cobalt chloride. After sintering the optical properties of the glasses produced were studied using UV/VIS-spectroscopy. Hereby, only glasses from suspensions with an amount of 0, 0.1, 0.2 and 0.4 wt.% cobalt chloride could be analyzed. Glasses obtained from suspensions with 0.7 and 1 wt.% were deep blue so that measurements did not lead to reasonable results. In Figure 7, the transmittance of sintered silica glass co-deposited from suspensions with different amounts of cobalt chloride is shown. The spectra are compared to that of a glass commercial fabricated.

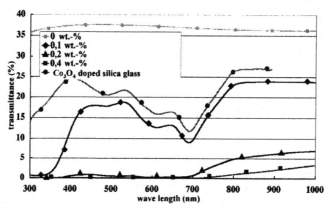

Figure 7: UV/VIS-spectra of different cobalt chloride doped silica

A parameter that limits the maximum amount of boric acid or cobalt chloride in the suspension is the electrical conductivity. Electrophoretic deposition of SiO_2 from suspensions with a value of σ higher than 20 mS/cm was not possible. If conductivity is too high the suspension will be heated to boiling and bubbles were deposited resulting in pores in the green body. Beside suspensions conductivity, the conductivity of the water with TMAH in the second chamber is also very important. It was determined that the conductivity in the second chamber had to be 10-fold of the suspensions conductivity. Due to the co-deposition of the ions, we achieved a doped green body that was more homogeneous compared to other methods such as gas infiltration, soaking of green bodies or by means of powder mixtures. Because of using nano-sized powders there were short paths of diffusion.

CONCLUSIONS

It was shown that manufacturing of doped of silica glass in a single-step process using EPD was possible. Boric acid and cobalt chloride were used as a model system to analyze if anions as well as cations could be co-deposited with nano-sized particles. Since boron oxide decreased the sintering temperature, the quantity was analyzed qualitatively by dilatometry combined by measuring the sintering temperature. The quantity of cobalt chloride in the glasses was appraised by the coloration. The amount of dopants in the green bodies was determined by ICP whereas it was shown that the quantity in the green bodies was twice larger than that in the suspension. The electrical conductivity in the suspensions must be smaller than 20 mS/cm in order to make deposition possible. Deposition with a ratio in conductivity between suspension and the second chamber about 1:10 resulted in better green density and green strength then for other ratios.

LITERATURE

[1] J. Tabellion and R. Clasen, "Advanced Ceramic or Glass Components and Composites by Electrophoretic Deposition/ Dmpregnation using Nanosized Particles", pp. 617-627, H.-T. Lin, M. Singh, Eds., 26th Annual Conference on Composites, Advanced Ceramics, Materials and Structures (The American Ceramic Society, Cocoa Beach, Florida, USA, 2002).

[2] E. Hückel, "Die Kataphorese der Kugel", *Physik. Z.*, **25** 204-210 (1924).

[3] J. Tabellion, R. Clasen, J. Reinshagen, R. Oberacker and M. J. Hoffmann, "Correlation between Structure and Rheological Properties of Suspension of Nanosized Powders", *Key Engineering Materials*, **206-213** 139-142 (2002).

[4] R. Clasen, "Verfahren zur Herstellung von dotierten Gläsern", *Fortschrittsber. Dtsch. Keram. Ges.*, **17** [1] 118-126 (2002).

[5] R. Moreno and B. Ferrari, "Advanced Ceramics via EPD of Aqueous Slurries", *Am. Ceram. Soc. Bull.*, **Jan 2000** 44-48 (2000).

[6] R. Clasen, "Forming of Compacts of Submicron Silica Particles by Electrophoretic Deposition", pp. 633-640, H. Hausner, G. L. Messing, S. Hiranos, Eds., 2nd Int. Conf. on Powder Processing Science (Deutsche Keramische Gesellschaft, Köln, Berchtesgaden, Okt. 12.-14. 1988, (1988).

[7] J. Tabellion and R. Clasen, "Near-Shape Manufacturing of Complex Silica Glasses by Electrophoretic Deposition of Mixtures of Nanosized and Coarser Particles", 28th International Cocoa Beach Conference on Advanced Ceramics and Composites (The American Ceramic Society, Cocoa Beach, Florida, (2004).

[8] J. Tabellion and R. Clasen, "Electrophoretic Deposition from Aqueous Suspensions for Near-shape Manufacturing of Advanced Ceramics and Glasses - Applications", *J. Mater. Sci.*, **39** 803-811 (2004).

[9] O. Stern, "Zur Theorie der elektrolytischen Doppelschicht", *Z. Elektrochem.*, **30** 508-516 (1924).

SHAPING OF BULK GLASSES AND CERAMICS WITH NANOSIZED PARTICLES

Jan Tabellion
Institute of Microsystem Technology
Laboratory for Materials Process Technology
University of Freiburg
Georges-Köhler-Allee 102
D 79110 Freiburg, Germany

Rolf Clasen
Department of Powder Technology
Saarland University, Geb. 43
D 66123 Saarbrücken, Germany

ABSTRACT

Manufacturing of functional ceramics and high-performance glasses by shaping of nano-particles and subsequent sintering combines significant advantages. Due to the high specific surface area of nanoparticles significantly increased sintering activity is achieved, which results in much lower sintering temperature. However, most of the common shaping techniques are not adapted to the intrinsic properties of nano particles. Due to their high specific surface area and low bulk density dry pressing can not provide an economic alternative. Suspension-based techniques seem to be much more promising to achieve green bodies with high density and good homogeneity. Nevertheless, with slip or pressure casting only comparably low compaction rates can be achieved, decreasing with particle size. In contrast, deposition rate is independent of particle size in case of electrophoretic deposition (EPD). Thus, EPD from aqueous suspensions is a fast and economic shaping technique for nanosized particles. A deposition rate of up to 3.5 mm/min was achieved, controlled mainly by the applied electric field strength (typically 1 to 10 V/cm). However, due to the high specific surface area of nano-particles, the density of the electrophoretically deposited green bodies was limited to about 50 %TD. By combining nanosized with larger particles, significantly higher green densities could be reached. From a suspension with optimized properties and adjusted parameters during EPD a green density of up to 81 %TD could be reached, resulting in a strong decrease in sintering shrinkage down to app. 6 %. Samples are shown for silica glasses and zirconia.

INTRODUCTION

In recent years, an increasing interest exists in using nanosized particles not only as fillers or functional isolated particles within a structural matrix but also for the manufacturing of mono-lithic glass and ceramic components. Due to their comparably very high specific surface area nano-particles bring a strongly enhanced sintering activity about. Thus sintering temperatures are reduced significantly compared to conventional microsized powders. Fully dense ceramics can be achieved avoiding high temperature phase transformations e.g. in case of zirconia[1]. Further-more, changed mechanical and physical properties can be achieved due to the high ratio of grain boundary to grain bulk[2]. In case of powder manufacturing of glasses, the use nano-particles is mandatory to avoid crystallization during sintering[3].

Manufacturing of large glass or ceramic components with high wall thicknesses from nanosized particles is accompanied by some severe disadvantages. First of all, preparation of nano-particles, which are appropriate to the manufacturing of ceramics and glasses on an industrial scale, is still cost-intensive, resulting in high cost of raw material. Furthermore, shaping of nano-particles is difficult, because most of the common techniques for advanced ceramics are not adaptable to the intrinsic requirements arising from the use of nano-particles. Both bulk density of nano-particles and green density after dry pressing are very low. The application of suspension-based methods seems therefore much more promising. However, for most of these techniques deposition rates depend strongly on particle size. In case of slip or pressure casting e.g., deposition rate decreases significantly with decreasing pore radius of the generated green body, which in turn is mainly governed by particle size of the powder used.

In contrast, deposition rate is independent of particle size in case of electrophoretic deposition (EPD)[4]. Thus a fast compaction of nano-particles can be achieved. Electrophoretic deposition is a shaping process where the driving force arises due to an externally applied electric field. Charged particles are forced to move through a dispersing medium towards an oppositely charged electrode by the electric field. On an oppositely charged electrode or on a porous dielectric mould arranged perpendicular to the moving direction of the particles between the electrodes the particles coagulate and form s stable deposit. Reviews about EPD as a shaping technique for advanced ceramics and glasses are given in references[5, 6]. Bulk ceramics shaped from nano-particles by EPD include alumina[7] and zirconia[8]. Furthermore, EPD has been used for the manufacture of transparent silica glasses. Thus, silica green bodies with a thickness of 10 mm could be shaped within 5 minutes and sintered to transparent glass at 1450 °C from nanosized Aerosil OX50[9]. Higher green densities were achieved from identical suspensions by means of EPD compared to slip casting or pressure casting[10]. Even for comparable green densities, a lower sintering temperature was observed for electrophoretically deposited silica green bodies, which is related to the higher microstructural homogeneity[11].

The aim of this work was to highlight the potential of electrophoretic deposition of both nanosized particles and powder mixtures of nanosized and conventional larger particles for the manufacture of bulk glass and ceramic components.

EXPERIMENTAL

Two kinds of both silica and zirconia particles were used, one nanosized powder and another one with a mean particle size of several microns. The silica powders used were Aerosil OX50 (Degussa) with a mean particle size of 40 nm and Excelica SE15 (Tokuyama, $D_{50} = 15 \mu m$). Furthermore, tetragonally stabilized zirconia powder with a D_{50} of 0.6 μm (Tosoh, TZ3YS) and a nanosized zirconia powder (Degussa Zirconia-3YSZ, mean particle size 30 nm) were used. Aqueous suspensions of OX50, mixtures of OX50 and SE15 and a mixture of 3YSZ and TZ3YS were prepared by dispersing the particles gradually in bidistilled water by means of a dissolver (LDV1, PC Laborsysteme), adding different amounts of tetramethylammoniumhydroxide (TMAH). Vacuum was applied to avoid incorporation of air bubbles. TMAH was used to adjust pH and thus the ζ-potential of the silica particles as well as the viscosity of the suspension, both of which are important factors concerning EPD. The solids content of the suspensions was varied between 10 and 60 vol. %, depending on the powders used.

Electrophoretic deposition was carried out under constant applied voltage by the membrane method[12]. A porous ion-permeable membrane was used as deposition surface. The electrophoresis cell was subdivided by this polymer mould in two chambers, so that deposition of particles (onto the porous mould) and recombination of ions (at the electrodes) were separated. Thus, no gas bubbles were found within the deposits. The cathode-sided chamber was filled with the suspension, the other with bidistilled water containing different amounts of TMAH. After determining the effective electric field strength within the electrophoresis cell, the width of the different chambers was adjusted, to allow for optimum deposition conditions[13]. A simple set-up with planar parallel electrodes and a planar polymer membrane was used to determine the influence of process parameters on deposition rate and green density. For more complex shaped components the set-up had to be re-adjusted concerning the shape of the porous mould as well as the electrode design. The applied voltage was varied between 1 and 15 V/cm. Deposition time was varied between 1 and 3 minutes.

After shaping, the green bodies were dried in air under ambient humidity. No cracking was observed. Sintering of the compacts was carried out either in vacuum (SiO_2 powder mixtures, 1480 °C), in air (zirconia, 1400 °C) or in a zone furnace (OX50, 1320 °C). Green density was measured by Archimedes method. The high depth of focus on the transmission micrograph was due to the software module EFI (Extended Focal Imaging by analySIS).

RESULT AND DISCUSSION

Concerning the manufacture of large bulk ceramics and glasses from nanoparticles, two key factors have to be considered. First of all, a deposition rate as high as possible is worthwhile and, secondly, a high green density in combination with good microstructural homogeneity is favorable in order to avoid high shrinkage during drying and sintering.

First of all the influence of the processing parameters on deposition rate was investigated. One important result is the fact, that shaping of green bodies from nanoparticles is outstandingly fast by means of EPD. As shown in Figure 1, for a suspension containing 31 vol.% OX50, which is about the highest solids content processable by EPD, a deposition rate of up to 30 g/cm²·min (dark, dotted line, y-axis on the left hand side) can be reached for an applied electric field strength of only 10 V/cm (in this case corresponding to an applied DC voltage of 30 V). With the resulting density of app. 40 % of the theoretical value (%TD) of 2,20 g/cm³, this corresponds to an growth rate of the green body of 3.5 mm/min. Furthermore, the deposition rate can be controlled by adjusting the electric field strength, allowing for a reproducible tailoring of the wall thickness of the green body. Thus, shaping of nanoparticles can be carried out very fast by means of EPD. The influence of other process parameters of the electrophoretic deposition on deposition rate was also investigated in detail for nanosized OX50 and is described elsewhere[14].

In comparison, the deposition rate for a suspension with 58 vol.% OX50 and SE15 (mixing ratio 10/90) is shown in Figure 1 (dark line). A slightly higher deposition rate was observed for the powder mixture, which was due to the higher solids content. However, in general, no influence of particle size on deposition rate was observed. Due to the significantly higher density (≈ 81 %TD) of the green body consisting of the mixture of SE15 and OX50 a distinctly lower growth rate of 2 mm/min was observed in this case.

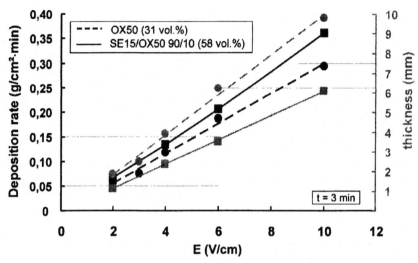

Fig.1: Deposition rate for EPD of nanosized fumed silica OX50 and a bimodal mixture of OX50 and SE15

In Fig. 2 the linear sintering shrinkage of numerous silica green bodies shaped EPD under different parameters and from suspensions with different solids contents is shown as a function of their green density. The dark dots represent samples shaped solely from OX50. As can be seen in Fig. 2, the green density of the OX50-samples is limited to values below app. 50 %TD. In comparison with other techniques this is still high[10]. The highest green density of 49.8 %TD was found for the sample deposited from the suspension with the highest solids loading (31 vol.-%). A further increase in solids content was not possible due to a strong increase in viscosity[15]., which made shaping by electrophoretic deposition impossible. As a result, linear shrinkage of more than 20 % occurs during sintering the OX50 green bodies at 1320 °C, to fully dense and transparent silica glasses. Apart from the typically high cost of raw material, this is the most significant disadvantage of using nanoparticles for the manufacturing of bulk ceramics and glasses. High shrinkage appears along with the risk of distortion, and near-shaping of complex geometries is impossible. Furthermore, the necessary oversize of the sintering furnace adversely affects the process cost.

Nevertheless, to benefit from the advantageous properties of nanomaterials, like the high sintering activity or the small resulting grain size after sintering, these disadvantages can be avoided by combining nano-particles with conventional, larger particles. As a result of an optimized ratio of particle sizes and an adjusted mixing ratio of nanosized and bigger particles, green bodies with very high green densities could be shaped. A very good microstructural homogeneity of the green bodies could be observed after drying. No size-dependent particle separation occurred. The linear sintering shrinkage of numerous silica green bodies shaped under different conditions by EPD from mixtures of SE15 and OX50 is also shown in Fig. 2 as function of green density (light rhombi). Much higher green densities could be achieved with these powder mixtures with a maximum of 81.7 %TD. This value was achieved for a green body shaped from a suspension with 58 vol.% OX50 and SE15 in a mixing ratio of 10/90. The resulting linear shrinkage during sintering to a fully dense and transparent glass at 1480°C was observed to be only 6 %.

Thus, two significant disadvantages of the use of nanoparticles for the manufacturing of bulk components can be overcome by using mixtures of nanosized and larger conventional microsized particles. Only ten percent of nanoparticles are necessary to achieve such high green density. Thus the cost of raw material can be reduced significantly. Furthermore, near-shape manufacturing of complex-shaped silica glass components is possible.

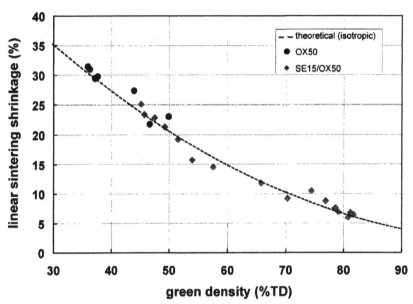

Fig.2: Linear sintering shrinkage of silica green bodies electrophoretically deposited from different OX50-suspensions (dark dots) and different suspensions containing bimodal mixtures of OX50 and SE15 (light dots)

Figure 3 shows a sintered silica glass triple mirror array as received after electrophoretic deposition and sintering. The corresponding green body was deposited from a suspension containing 58 vol.% OX50 and SE15 in a mixing ratio of 10/90. Sintering was carried out at 1480 °C. Due to the high green density only very low shrinkage occurred during sintering as shown above (cp. Figure 2). Since the shrinkage occurred isotropically, no deformation of the structures was achieved, as can be seen on the transmission micrograph on the right-hand side of Fig. 3. Sharp edges and defined contours were achieved without any post-processing, which is not possible in case of silica glass by conventional manufacturing via the melting route.

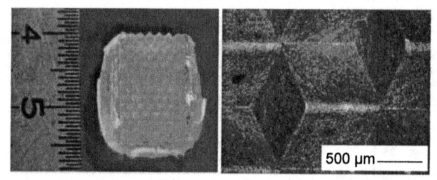

Fig.3: Silica glass microstructure as received after EPD (bimodal mixture of OX50 and SE15) and sintering (right hand side: transmission micrograph with software module EFI)

A similar optimization process was carried out in case of zirconia, but will not be described in detail here. Figure 4 shows a sintered zirconia tube with an inner diameter of 7 mm and a wall thickness of 1.3 mm after sintering at 1600 °C. Isotropic shrinkage of only 12 % (linear) occurred during sintering.

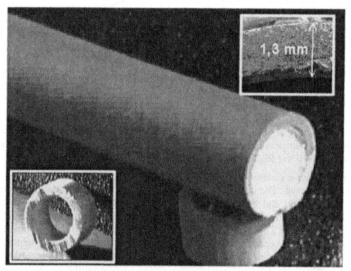

Fig.4: Zirconia tube (inner diameter 7 mm, wall thickness 1.3 mm) shaped by EPD of a bimodal powder mixture, as received after sintering (1600°C)

CONCLUSIONS

Using nanoparticles for the manufacturing of bulk ceramic and glass components can result in components with enhanced properties. However, still the high cost of raw materials as well as the typically low density of green bodies from nanoparticles are getting in the way of a large-scale production of bulk components from nanosized particles. Although very high deposition rates can be reached by electrophoretic deposition of nano-particles (up to 3.5 mm/min for nanosized fumed silica OX50) even for a very low energy input (E < 10 V/cm, t <3 min), the green density is limited to about 50 %TD. Combining nanosized particles with conventional particles, having a mean size of several microns, can result in green bodies with significantly enhanced density. Due to the fact that the deposition rate is independent of particle size in case of EPD, very homogeneous green bodies were achieved, showing no size-dependent particle separation. A maximum green density of app. 81 %TD was reached. Correspondingly, the sintering shrinkage could be reduced significantly (app. 6 % linear). Thus near-shape manufacturing of complex shaped silica glasses or zirconia components is possible, allowing at the same time for a much more cost effective production of bulk ceramics and glasses from nano-powders, because much less nano-particles are necessary to achieve similar favorable materials properties.

REFERENCES

[1]M. J. Mayo, "Processing of Nanocrystalline Ceramics from Ultrafine Particles," *Int. Mater. Rev.*, **41**, 85-115 (1996).

[2]H. Hahn and K. A. Padmanabhan, "Mechanical Response of Nanostructured Materials," *Nanostructured Materials*, **6**, 191-200 (1995).

[3]E. Rabinovich, "Review: Preparation of Glass by Sintering," *J. Mater. Sci.*, **20**, 4259-4297 (1985).

[4]C. Hamaker, "Formation of a Deposit by Electrophoresis", *Trans. Faraday Soc.*, **36**, 279-287 (1940).

[5]M. S. J. Gani, "Electrophoretic Deposition - A Review," *Industrial Ceramics*, **14 [4]**, 163-174 (1994).

[6]P. Sarkar and P. S. Nicholson, "Electrophoretic Deposition (EPD): Mechanisms, Kinetics, and Application to Ceramics," *J. Am. Ceram. Soc.*, **79 [8]**, 1987-2001 (1996).

[7]B. Ferrari and R. Moreno, "Electrophoretic Deposition of Aqueous Alumina Slips," *J. Eur. Ceram. Soc.*, **17**, 549-556 (1996).

[8]K. Moritz, R. Thauer, E. Müller, "Electrophoretic Deposition of Nano-scaled Zirconia Powders Prepared by Laser Evaporation," *cfi/Ber*, **77 [8]**, E8-E14 (2000).

[9]R. Clasen, "Preparation and Sintering of High-Density Green Bodies to High Purity Silica Glasses," *J. Non-Cryst. Solids*, **89**, 335-344 (1987).

[10]J. Tabellion; R. Clasen, "Electrophoretic Deposition from Aqueous Suspensions for Near-shape Manufacturing of Advanced Ceramics and Glasses—Applications", *J. Mater. Sci.*, **39 [3]**, 803-811 (2004).

[11]J. Tabellion, E. Jungblut and R. Clasen, "Near-Shape Manufacturing of Ceramics and Glasses by Electrophoretic Deposition Using Nanosized Powders"; in *Ceramic Engineering and Science Proceedings*, **24 [3]**, AcerS (Westerville), 75-80 (2003).

[12]R. Clasen, "Forming of Compacts of Submicron Silica Particles by Electrophoretic Deposition", in *2nd Int. Conf. on Powder Processing Science*, ed. H. Hausner, G. L. Messing and S. Hirano, Berchtesgaden,: DKG,. 633-640 (1988).

[13]J. Tabellion and R. Clasen, "In-Situ Characterization of the Electrophoretic Deposition process"; in *Innovative Processing and Synthesis of Ceramics, Glasses, and Composites IV.,* ed. by N. P. Bansal and J. P. Singh. Am. Ceram. Soc., Westerville, 197-208 (2000).

[14]J. Tabellion and R. Clasen, "Controlling of Green Density and Pore Size Distribution of Electrophoretically Deposited Green Bodies"; in *Innovative Processing and Synthesis of Ceramics, Glasses, and Composites IV.,* Edited by N. P. Bansal and J. P. Singh. Am. Ceram. Soc., Westerville, 185-196 (2000).

[15]J. Tabellion, R. Clasen, J. Reinshagen, R. Oberacker and M. J. Hoffmann, "Correlation Between Suspension Structure and Rheological Properties of Suspensions of Nanosized Fumed Silica Powders", in *Improved Ceramics through New Measurements, Processing, and Standards*, ed. M. Matsui, S. Jahanmir, H. Mostaghaci, M. Naito, K. Uematsu, R. Wäsche and R. Morell, ACerS, Westerville (USA), 183-188 (2002).

Author Index

Author Index